To the Brandywine River Museum

Enjoy the view!

Ben Rappaport

June 2002

H

COLLECTING ANTIQUE MEERSCHAUM PIPES

MINIATURE TO MAJESTIC SCULPTURE

Ben Rapaport

Photography by Gary L. Kieffer/Photo Press International, Javier A. Flores, Stephen Crawley, and others.

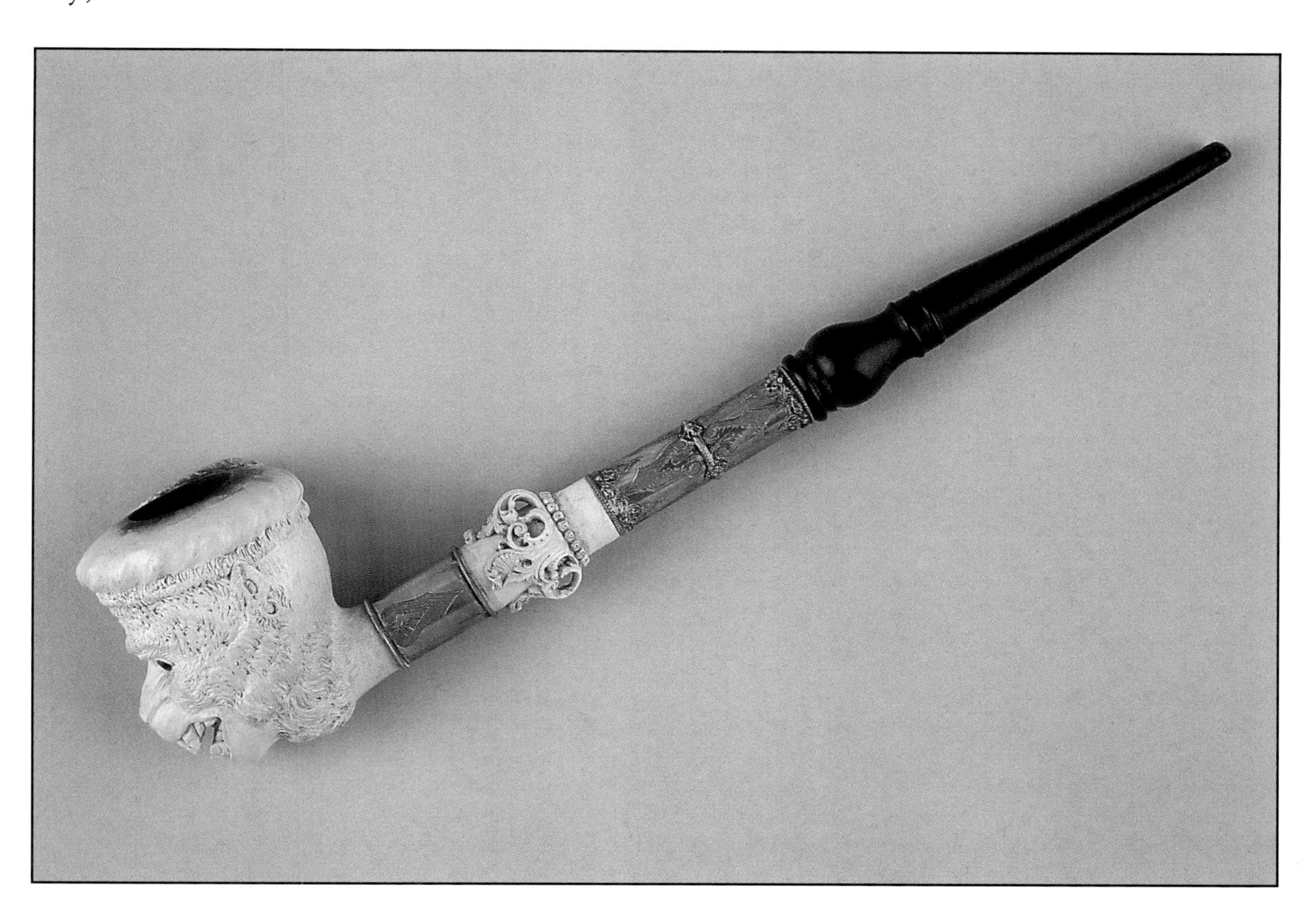

4880 Lower Valley Road, Atglen, PA 19310 USA

Cover photo: A commemorative cheroot holder depicting the capture of the Austrian cavalry standard by a Prussian officer and three dragoons at the Battle of Sadowa. Sadowa was once a village in Northeast Bohemia, Western Czechoslovakia, where Prussia, Italy, and some minor German states opposed Austria, Saxony, Hanover, and the states of Southern Germany in what was called the Seven Weeks' War of 1866. The holder measures 14" l., 4" h. prob. German, ca. 1875. According to company archives, Alfred Dunhill Sr., the founder of Alfred Dunhill Pipes Ltd., London, England, acquired this holder about 1920. *Courtesy of the Alfred Dunhill Archive Collection.*
Back cover photo: Detail of pipe allegedly made for the *Exposition Universelle,* Paris, 1889. The pipe measures 28" l., 11" h., prob. French, ca. 1888. (See page 59.) *Courtesy of the SP Collection.*

Library of Congress Cataloging-in-Publication Data

Rapaport, Ben.
Collecting antique meerschaums: miniature to majestic sculpture, 1850-1925/Ben Rapaport: photography by Gary L. Kieffer...
p. cm.
Includes bibliographical references.
ISBN 0-7643-0765-7
1. Meerschaum tobacco pipes--Collectors and collecting--History--19th century--Catalogs. 2. Meerschaum tobacco pipes--Collectors and collecting--History--20th century--Catalogs. I. Title.
NK6045.R36 1999
688'.42--dc21 98-44885
CIP

Designed by "Sue"
Type set in ZapfCalligr BT/Garamond

ISBN: 0-7643-0765-7
Printed in China

Published by Schiffer Publishing Ltd.
4880 Lower Valley Road
Atglen, PA 19310
Phone: (610) 593-1777; Fax: (610) 593-2002
E-mail: Schifferbk@aol.com
Please visit our web site catalog at **www.schifferbooks.com**

This book may be purchased from the publisher.
Include $3.95 for shipping.
Please try your bookstore first.
We are interested in hearing from authors
with book ideas on related subjects.
You may write for a free catalog.

In Europe, Schiffer books are distributed by
Bushwood Books
6 Marksbury Rd.
Kew Gardens
Surrey TW9 4JF England
Phone: 44 (0)181 392-8585; Fax: 44 (0)181 392-9876
E-mail: Bushwd@aol.com

Dedication

This book is dedicated to three distinct groups of people. First, and above the rest, is family—my wife, Liz, and my son, Darren. For countless years, they have tolerated my many absences on global treks, late-night visitors to our home, and long-distance phone calls from far-off places, an unreasonable price to pay for my commitment and devotion to a very arcane hobby.

To the second group, comprised of all those who love art regardless of the medium in which expressed, I ask that you examine this book. I am confident that if you take the time to look, you'll discover diminutive to prominent sculptured art at its very best.

The ancient Greek poet, Archilochaus, wrote an essay comparing the fox, which knew many things, and the hedgehog, which knew one big thing. If applied to this field, the fox would collect all manner of antique smoking artifacts, and the hedgehog would collect only meerschaums. To all the hedgehog pipe collectors and to every fox pipe collector, who, from time to time, has succumbed to the allure of an exquisitely carved antique meerschaum and added it to his collection, I owe a special vote of thanks for having encouraged me throughout this project, and for having donated their time and their prized possessions for exhibition herein. Of these all, the most persistent, unrelenting, indefatigable, and tenacious taskmaster was Dr. Sarunas "Sharkey" Peckus, who requested that I start ... and finish this book during his lifetime. Sharkey, your wish was my command, and your wish has been realized! As well, I owe a very special debt of gratitude to David R. Wright, the curator of the Museum of Tobacco Art and History in Nashville, Tennessee, from 1982 to 1998. He was always very supportive of us and anyone who took an interest in tobacco collectibles. David, we will miss you and the Museum!

May all the foxes and all the hedgehogs of this hobby always find the beauty—and never the beast—in their continuing quest for these *objets d'art.*

The neoclassical Three Graces—Aglaia (Brilliance), Euphrosyne (Joy), and Thalia (Bloom)—in Renaissance paintings by Botticelli and Raphael and a sculpture by Antonio Canova were captured in this carved-in-the round, hookah-style pipe by Ludwig Hartmann & Eidam, Wien, for the *Weltausstellung*, Vienna, 1873, 31.5" h. with pedestal. *Courtesy of the GC Collection.*

Contents

Foreword 7

Part I: Historical Background 9

Chapter I. Introduction 9

Chapter II. The Smoking Pipe As Art Form 10

Chapter III. Meerschaum, In the Beginning 11

Chapter IV. A Perspective on the Trade 19

Chapter V. Dating and Attribution: The Metamorphosis of the Meerschaum 31

Chapter VI. Expositions and Exhibitions Yesterday and Today 41

Part II: The Collector's Market Today 47

Chapter VII. Valuation: How Much Is It Worth? 47

Chapter VIII. Searching for Holy Grails 51

Part III: Meerschaum Retrospective 56

Chapter IX. Celebration and Commemoration 58

Chapter X. Femmes Fatales 66

Chapter XI. Of Mice and Men 84

Chapter XII. Myth and Fable 107

Chapter XIII. Naiads, Putti, Cherubs, and Seraphim 117

Chapter XIV. Whimsy, Fancy, Quaint, and Curious 128

Chapter XV. Naïve, Suggestive, and Erotic 145

Endnotes 153

Annex A. Makers and Marks 156

Annex B. International Exhibitions and Expositions, 1851-1901 164

Annex C. The Right Places for the Right Stuff 166

References 169

Price Guide 173

Index 175

Foreword

I have often heard that, ornamentation aside, all pipes are pretty much the same. If you hold this opinion about all smoking pipes, I suggest that you view the contents of this book with a discerning and diligent eye. Can an old smoking pipe be a work of art? I believe that the many masterpieces showcased in this volume answer the question with a resounding "yes!" Why a book just about meerschaums when, by definition, antique pipes include myriad mediums, natural and man-made, and national expressions from just about every corner of the globe? One answer is that I wrote a book about antique pipes 20 years ago.[1] A better answer is that although I collect all manner of antique pipes, I am partial to antique meerschaums and this particular era. However, my partiality should not be construed to imply that pipes of wood, porcelain, and clay of the same period were not of quality. In my opinion, all these mediums met the practical requirements of form, fit, and function for the pipe smoker, but the medium of meerschaum had added character not easily expressed in these other mediums—the finesse of two- and three-dimensional sculpture in a variety of possible expressions. Then, why not a book about the entire history of the meerschaum pipe? After all, the universally accepted legend about the discovery of meerschaum as a pipe medium is dated to the 1720s, and meerschaum pipes are still being produced. Why a book covering a period of only 75 years? My answer to these two questions is that, when compared with the earliest carved meerschaums of the 18th century and those produced in Turkey today, those crafted in the latter half of the 19th century to the first quarter of this century are simply far superior in execution and flair. Collectively, meerschaums of this period were the quintessential smoker's requisites.[2]

It is said that art imitates life, and antique meerschaums have imitated every aspect, dimension, and facet of life, but always in the diminutive, and almost always in exacting scaled proportion. A multitude of antique meerschaums are precise facsimiles of art, because the carver often copied the art of others, interpreting the creations of the sculptor, muralist, illustrator, etcher, and engraver, but always in his own medium. A large proportion of antique meerschaums was original and innovative art, imaginative expressions of some very inventive craftsmen. Yet another class of meerschaums is comprised of what I call facsimile art. Free trade and the absence of intellectual property laws gave license for one carver to copy the popular motifs of another, but almost always adding or subtracting a nuance here, a bit of flair there. Many antique meerschaums may look alike, or have a similar design or size, but there, the resemblance ends. Since the meerschaums of this era were never mass-produced, even the best impersonations were not ever exact copies.

It should be evident that I have a strong opinion about meerschaums and about this period, a period of very prolific carving in a number of capitols of Western Europe and a handful of cities in the United States. I am not alone in this view. In a respected newspaper of the period, the following appeared: "A meerschaum pipe is one of the best things a man can have, and gives him more pleasure for the amount of money expended than anything else he can buy."[3] What was available to buy about 100 to 150 years ago was an infinite variety of shapes and sizes; varying degrees of intricate bas- to high-relief-carved subject matter; assorted appliquéd embellishments and decoration, such as semi-precious and precious gems, and gold and silver *repoussé* accents; mouthpieces of amber, or ivory, or exotic woods; and custom-made fitted carrying cases to protect the fragile meerschaum from possible damage and shield it from the unforgiving sun or harsh showroom light that could dissipate the hand-rubbed beeswax finish. In a word, name something—almost anything—and chances are that it was expressed in meerschaum. Many are one-of-a-kind pieces of history in miniature, some qualify as architectural statements, and most evoke ingenious creativity and dramatic imagery. And the best compliment ever paid to the antique meerschaum is that it is as skillfully elaborated and as intricately fashioned as Japanese netsukes. That's why collecting meerschaums is so fascinating.

Something more. Collecting meerschaums is really a 20th century avocation. The pipes and cheroot and cigarette holders illustrated in this book were very popular as smoker's requisites during the last century, but there were very few serious collectors in Europe or in the United States during the period. History acknowledges only three prominent 19th century pipe collectors, and all three were European— Augustus Frederick (Queen Victoria's uncle, the Duke of Sussex and the Earl of Inverness) and William Bragge, F.S.A., of England, and the Baron Oscar de Watteville, of France. The most noted 20th century American pipe collector was John F. H. Heide of Chicago, Illinois. He began collecting in 1900, and continued his pursuit of specimens from around the world, but of the almost 1,400 pipes and related tobacco artifacts he had amassed, meerschaums were not found in significant numbers when his collection was auctioned in 1946. Since the 1950s, the number of collectors of antique meerschaum has grown considerably and, by my count, American collectors outnumber foreign collectors.

So, follow me, please. Allow me take you on a trip through yesteryear when men—and some women—used a meerschaum pipe or one of its diminutive counterparts, the cheroot holder or cigarette holder, for smoking pleasure. I invite every inveterate, neophyte, and would-be pipe collector to join me in a front-row seat on a first-class flight of fancy through (tobacco) time to have a glimpse at art long since gone. Look and you will understand why today's antique meerschaum collector is so frenetic about his hobby. The illustrations of a small fraction of what was produced in these 75 years vividly demonstrate why the meerschaum collector derives immense pleasure from the appearance and tactility of a humble utensil that was, a century ago, a mere vessel for tobacco. Today, these vessels are no longer prized for their smokeability; they are cherished, valued, and prized as objects of art. This book is bound to stimulate all the senses, because it contains images of an abundance of matchless, unprecedented, unique, uncommon, even nonpareil examples of the art of meerschaum carving during a fraction of the lengthy 300-year history of its use as a medium for smoker's utensils.

Part I: Historical Background

Chapter I. Introduction

Collecting Antique Meerschaums: Miniature to Majestic Sculpture, 1850-1925 is not an incisive research monograph about the history or evolution of this mineral. Little space is devoted to the origin of meerschaum, how these pipes were made, or how well they smoked. This is an art book, a catalog of exhibits illustrating the infinite majesty and magic of this substance which was originally crafted into utilitarian and mundane smoking utensils— pipes, cheroot (little cigar) holders, and cigarette holders. Some of the utensils illustrated in this book even surpass ordinary creativity and craftsmanship and verge upon artistic genius, to be compared with the larger-than-life creations in bronze, marble, alabaster, and stone of Brancusi, Cellini, Maillol, Michelangelo, Moore, Saint-Gaudens, and other noted sculptors. Sadly, the names of the artists to whom many of these mini-sculptures are attributed are little-known or unknown in the world of mainstream art, but their names are revered by meerschaum collectors. These artists fashioned and formed works of art from the aristocrat of smoking substances, commercially known as hydrous magnesium silicate, a member of the soapstone family.

Pipe-making, in general, has always been a rather arcane handicraft, best described as a cottage industry, and, unfortunately, shrouded in much mystery... and occasional myth. Too few facts are known, even fewer details were recorded for posterity, and later generations of most 19th century carvers recall almost nothing about those golden years of superior craftsmanship and artistry. No one in our collecting community has yet been able to satisfactorily demystify or untangle the too few recorded accounts of how and exactly when this small and select enterprise of craftsmen evolved, then burgeoned and, in time, eclipsed. For all my years of collecting, I am still not able to disentangle fact from fancy. I am also ill-equipped to clarify the obscure antecedents or speculate on what might have been the true nature of what could aptly be called an informal hanse, a loosely-knit league of a host of anonymous artisan-technicians. These trade craftsmen, whose lasting works amply demonstrate that they were also master practitioners deserve much more recognition than they collectively attained. They rightfully, should have earned their place in the annals of art history, but have remained unacknowledged until now.

This book will not answer the vast array of questions that typically arise in a committed collector's quest for perfect knowledge, answers to the six interrogatives of the inquisitive mind—who, what, when, where, how, and why—but it does offer a perspective on and an impression of yesteryear as others have recorded it. In the absence of factual evidence, quotations from primary sources are used to place the reader in the period, to permit him to envision a scenario of the times, the places, and the milieu of this cottage industry as it might have been. I personally believe that the pipes and holders appearing in this book are only a representative fraction of what was probably produced in the 75 years under study. I suspect that if it were possible to assemble and array in one place the output of just the known meerschaum pipe manufactories here and abroad, the harvest would, indeed, be bountiful and breathtaking. To comprehend more, please turn the page.

Chapter II. The Smoking Pipe As Art Form

A standard dictionary definition of a pipe is a device made of wood, clay, meerschaum, hard rubber, or other material, with a small bowl at one end, used for smoking tobacco. This definition excludes the descriptive word art. Reaching back into 19th century literature, I found a rather odd definition of a pipe. The author postulated 11 scholastic definitions of a pipe's various parts, i.e., bowl, stem, lid, mouthpiece, etc., its essential and accidental colors, and its simplicity and complexity. Scholastic Definition VII is the ornament: "when a pipe, be it in the bowl or be it in the stem, hath not a flat surface."[4] This definition hints that the pipe can be an art form. Interestingly, ornament, in this sense, acknowledges only variations in planar shape. I turned to another treatise on tobacco pipes to find "that modern Pipe par excellence—the Meerschaum. It has a technique and a vocabulary of its own,—not rendered the more intelligible in that they are preserved chiefly in foreign treatises of great erudition."[5] Searching for yet another definition of a meerschaum pipe, I found: "pipes in which the bowls are carved from soft, porous, generally white meerschaum. They are light in weight, have a high absorptive capacity, and are fragile. The mouthpieces of such pipes are usually made of genuine amber though in cheaper ranges they are sometimes of hard rubber."[6] This might serve as an appropriate definition for the tobacco trade, but it, too, lacks descriptive terms such as art or craft.

In the absence of a more fitting definition, and to characterize the subject matter in this book, suffice to say that regardless of how writers of that day defined the smoking pipe—and more specifically, the meerschaum pipe—the utensils of that period were distinctively individual. Inspiration and ideas for the carver came from everywhere: victory and defeat in battle; the birth, marriage, and death of royalty; mythology; heroes and heroines; the hunt; and even everyday life. Literature, paintings, and opera, and their authors, artists, and composers, respectively, were opportune motifs for the carver. At times, the carver responded to a special request. Left alone to cogitate, the carver was limited only by his own imagination, creativity, and skill. Every piece illustrated herein was made to be smoked, not necessarily envisioned to be savored and relished as a future art form. You, the viewer must decide if these creative objects are true art, because art, like beauty, is in the eye of the beholder.

Chapter III: Meerschaum, in the Beginning

Meerschaum—sea foam in German—or foam clay (*Spuma marina seu Taecum lithomarga,* in Latin) has a rather exotic and enchanting ring.[7] It is a poetic figure of speech, because meerschaum has been described as the aristocracy among smoking pipes; the queen of pipes; soft and light as a fleeting dream; creamy, delicate, and sweet as the complexion of young maidenhood; one of the choicest and rarest gifts of the gods; and the "apple of the eye" of the refined pipe smoker. It has also been called Venus of the Sea, White Goddess, or as defined by King Media, "Farnoo, or the froth-of-the-sea, an unctuous, argillaceous substance." [8] Parenthetically, this German word, in customary use for more than 250 years, has never been translated into English and it has no English-language equivalent!

It is not essential to detail or understand its mineral origins to better appreciate the contents of this book. It is enough to know that meerschaum is a heavy, fat, soft, mineral which can be worked almost as easily as clay. It is hydrated silicate of magnesia, or a compound of magnesia, flint, and water, in variable proportions. For the more curious and inquisitive reader, meerschaum is composed of 40.25 magnesia, 45.00 gravel soil, 10.14 carbonic acid and water, and 4.61 pipe clay. In its natural state, it may be found in colors from pale yellow to deep brown, contingent on the amount of proximate silicate of iron. Although meerschaum is found in many places in the world, the largest deposits of this mineral are found in Turkey. It is a substance that has, at times, been pressed in molds to form flowers, beads, and ornaments. Its principal use for more than 250 years, however, has been almost exclusively for the production of smoker's articles.

Cutting amber at the B. Linger Company, Vienna, Austria, ca. 1900.

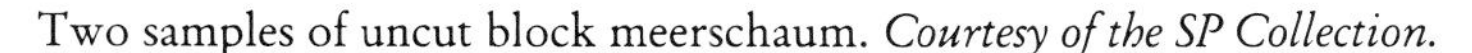

Two samples of uncut block meerschaum. *Courtesy of the SP Collection.*

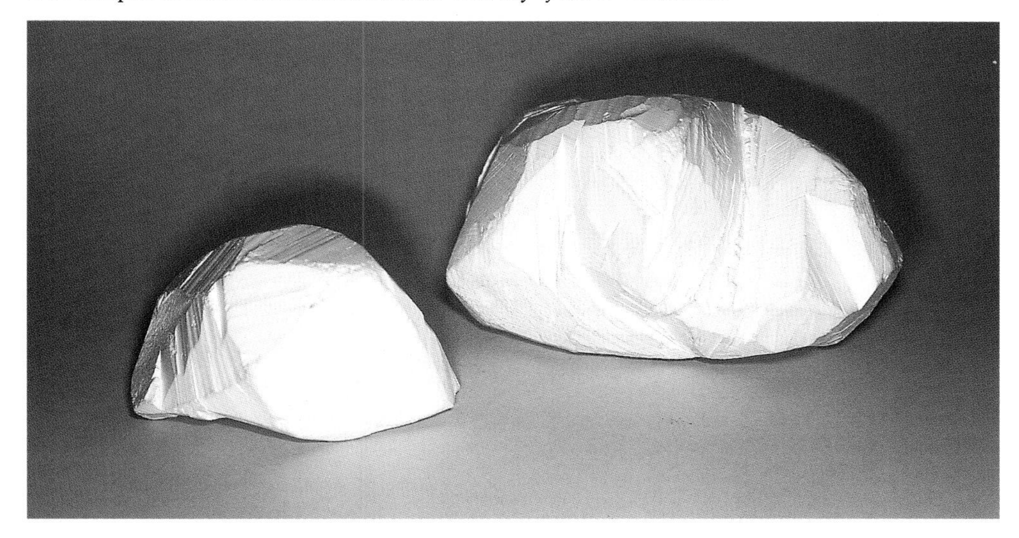

View of the amber turning room at the B. Linger Company, Vienna, Austria, ca. 1900.

BLOCK MEERSCHAUM. Early writers suggested that a certain German chemist, Johann Wiegleb, was the first to study the mineral in the 1790s, and he was to have decomposed meerschaum into various densities. In its raw natural state, pure meerschaum:

> shows a great difference in the substance, but a still greater one in the shape; and it can safely be said, amongst 100,000 pieces, no two will be found alike; the difference in the substance is in the weight, color, and hardness; some are light, others medium heavy, some as heavy as stone; some are soft, some get harder through smoking, some softer and more spongy. Some have a splendid white, some a more grayish color, even brown spots, some have hard and soft spots, cloudy-like veins, etc., etc., etc. These so-called veins, which are mostly found in the harder grades of Meerschaum, have been created through lengths of time. The surface of the Meerschaum fields generally consists of gravel, and the deeper the Meerschaum lays, the softer, lighter, and less stony it will be. The veins arise, however, through the ever working nature of the Meerschaum, and the less the Meerschaum is heavy and stony, the less they are visible; this is very important to the Meerschaum workers, especially when making plain goods, as on figured articles it can generally be cut out, for a good workman can see the veins on the surface of the raw Meerschaum, and will arrange his figure accordingly.[9]

A 1905 Kapp & Peterson pipe catalog counseled the meerschaum pipe smoker to be aware of three different grades of block meerschaum—hard, medium and soft—and suggested that a "wet" smoker should purchase a hard quality meerschaum pipe which would be very absorbent; the medium weight meerschaum pipe was suited for the average smoker; and the "dry" smoker should select the softest quality.

IMITATION MEERSCHAUM: In contrast to genuine meerschaum, imitation meerschaum—variously advertised as mock meerschaum, mere-sham, meerschaum chips, compressed meerschaum, meerschaum calcinate, meerschaum-masse, and Viennese meerschaum (neither Vienna nor the rest of Austria had mine deposits)— was described as having been waxed and processed as genuine meerschaum, had the same appearance and would color "like the real article." Not quite so! According to another Zorn Company catalog, imitation meerschaum is:

> an article manufactured from the chips and small pieces of the genuine material. They are first ground or mashed into a pulp, and then mixed with some binding chemicals, so as to keep the molecules together. It is then pressed into the shape intended, but left somewhat larger; after that it is dried and shrinks down to the size required. It is then worked on the lathe and finished the same as real Meerschaum...A few of the greatest differences between the two are as follows:—The appearance of the material is harder; there is no depth or life to it. Then, it is impossible to make any compound without getting some dust into it, and with a sharp eye or a magnifier these particles of dust can be detected, even in the finest imitation that is made.[10]

Henley's Twentieth Century Book of Formulas, Processes and Trade Secrets, a standard reference work on formulas stated that, to distinguish imitation meerschaum from genuine block meerschaum, one need only rub it with silver. "If the silver leaves lead pencil-like marks on the mass, it is not genuine but artificial meerschaum."[11] Others have claimed that it is easy to distinguish between block and imitation: the latter is considerably heavier by comparison, owing to the absence of foreign minerals, it is usually of a more even texture, and it does not take color as well as the natural mineral. Alternative artificial mixtures were also engineered that would not respond to this test because their density and composition responded yet differently.

CARVING: Each pipe-carver of that day who worked in porcelain, clay, wood, ivory, and other materials, used a slightly different set of pipe tools and applied a different set of skills to his respective trade. If he were to carve meerschaum, he would have required special instruction on how to attack a lump of soft-paste meerschaum. The knife was the most important implement of the carver's bench tools, and next in importance was the turning-lathe. Using all the special tools of this trade (which one author considered to be the paraphernalia of a torture chamber), the carver turned, cut, bored, polished, soaked in wax and tallow, repolished, and carved this material with more or less intricacy, and then fitted the finished pipe with an amber mouthpiece. Cutting, boring and, finally, polishing a meerschaum pipe were precise and intricate operations, which ultimately separated the amateur- and the connoisseur-carver. As one writer described the last step in the finishing process, the pipe was rubbed "with glass-paper, boiled in wax, spermaceti, or stearine, and polished with bone-ashes or chalk."[12] And the fin-

View of the pipe fitting room at the B. Linger Company, Vienna, Austria, ca. 1900.

View of the silver mounting workroom at the B. Linger Company, Vienna, Austria, ca. 1900.

ished product, typically long, thin, and curvilinear, required very delicate treatment, a markedly different treatment from that used in making a pipe from a briar ebauchon.

COLORATION: Coloration, or more accurately, pre-coloration, was a process invented to infuse a uniform color ranging from ivory to that of dried tobacco leaf onto the entire surface of the meerschaum, giving it the appearance of having been well smoked. One reason for applying such finishes, in the opinion of tradesmen, was that few smokers, even those who delighted in coloring a meerschaum pipe by smoking it naturally, ever succeeded in bringing it to a perfect and uniform color which, through use, could range from honey to tortoise-shell brown. Other than wax or oil, a mixture of dragon's blood (the inpissated juices of various plants of the genus Dracaena) and nut oil, had also been suggested; when the meerschaum thoroughly absorbed this mixture, the pipe turned a blackish-red hue. Many after-sale home remedies were offered, including the following:

- Clean the interior and exterior with alcohol, remove the mouthpiece, and bake the bowl in an oven at 130° to 140° to remove the wax and superfluous tobacco juices (length of time is unspecified).
- Remove scratches and dents with linseed oil and No. 0 pumice stone rubbed smoothly with a piece of soft flannel. Then rub the surface with oxalic acid, set aside for three to four hours, then rubbed lightly again.
- Make a paste from oxblood and Soudan brown analine dyes mixed with linseed oil and rub this mixture smoothly over the entire surface of the bowl. Burn this mixture in over an alcohol lamp.
- Cool the pipe, and polish it with rottenstone and water applied with a piece of cotton flannel.[13]

The smoker could have followed another, less complicated, alternative that was offered to the public:

> Those desiring to have their Meerschaum pipes and cigar holders to color beautifully, must be careful to avoid handling them with their fingers while warm from smoking, as the spots produced thereby are difficult to remove, and frequently cannot be removed at all. They should also never put them in the case until perfectly cool, and in laying them down to cool off, place them on blotting-paper, or what is preferable, put them on a pro-

jecting wire, that will fit the hole of the tube or the mouthpiece, and leave it free from contact with anything while cooling. Meerschaum being both hard and soft, the latter absorbing more wax than the former during the process of finishing, the soft pipe bowls require more heat in the start than the hard ones, in order to give them a fine, uniform, chestnut brown color. Cigar holders, on the contrary, should never be allowed to get warm, at least not more than milk-warm, as they should begin to color from the outer end. Care must be taken that new pieces are not smoked in the cold or draft, as it is very disadvantageous to the Meerschaum, causing so-called air-cracks, which never disappear; this care is absolutely necessary for the first 7 or 8 days. Further: New Meerschaum bowls must be smoked out to the bottom in order to give them a uniform heat, as if this is not done, the wax will not withdraw uniformly, and therefore cause an even color. Meerschaum bowls and pipes should never be covered with leather or cloth, as was fashionable some fifteen or twenty years ago, as it results in great disadvantages: *first*, the pipes cannot evaporate enough; *second*, the wax, which melts through the heat, sticks to the leather or cloth, hardens, and gives the article a muddy brownish color. Meerschaum pipes and tubes should not be wiped with cloth or handkerchief, as it is the habit with many smokers, but with a clean, soft piece of silk or chamois, to avoid all scratches. The inside crust, forming in pipes and bowls, should never be scraped or cut out with a knife, as the crust is an advantage to the coloring of the pipes, and only becomes dangerous if the crust gets too thick, when it may crack or burst the pipe; in the latter case, it should be sent to the experienced pipe turner or repairer, who can remove it properly and safely.[14]

All these procedures were very complex, and there is ample evidence that many tried and failed to color their pipes and holders this uniform brown hue, since just about every collector today owns a few mottled, speckled, or streaked brown antique meerschaums.

This figurative head of a St. Bernard is an example of a pre-colored pipe. No smoker could have achieved this uniform coloration through smoking, however much and however often he may have tried. The pipe is 7.75" l., 3" h., silver mount engraved "P.B.S. to J.G.B., 1894," prob. American. *Courtesy of the AZ Collection.*

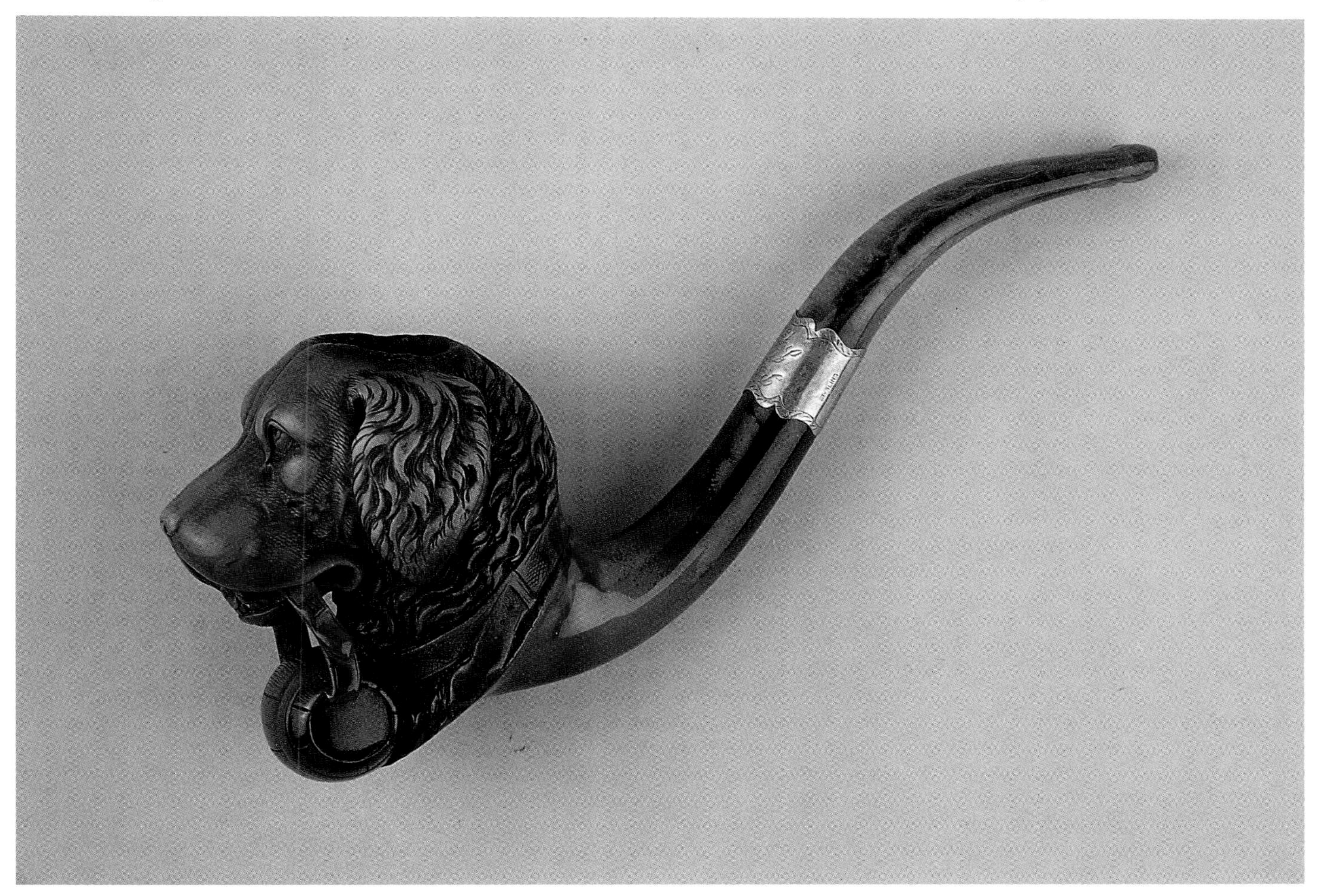

DISCOVERY. One source dates the word meerschaum to 1475, and it could be conjectured that the substance was known long before that. It is also possible that it was much later when the word was applied to or associated with the utensils for smoking tobacco. John III Sobieski (1624-1696) fought against the Tartars and the Cossacks in 1651-1652; as commander-in-chief of the Polish Army, he defeated the Turks at Hotin in 1673. Elected as the King of Poland in 1674, he later led the army against the Turkish siege of Vienna in 1683. A soft white substance was honorably mentioned in one account of King John's daring rescue of that city. Could King John have been an early witness to Turkish craftsmen shaping meerschaum into utensils for the pipe smoker?

The mottled or speckled finish throughout is the result of the smoker's inexperience. The cuffed hand is holding what appears to be a chess piece (?), the bust of a young jockey. Those who collect briars would call this pipe a setter because of its flat bottom. The pipe is 6" l., 3.5" h., prob. English or American, ca. 1890. *Courtesy of the Author's Collection.*

The universally accepted story of the first use of meerschaum for a pipe is found in just about every book on pipe smoking. It has been abundantly recorded that a certain cobbler, Karl Kovács (or Karol Kowates and other spellings), of Pest, Hungary, was the first to have made two pipes in 1723, one for himself and one for a Hungarian count named Andrássy who gave him two lumps of raw meerschaum; the count had received the blocks as a gift from Sultan Achmed III of Turkey. One recent version reads:

> Kovács Károly (woodcarver, inventor of the meerschaum pipe, born in Hungary) lived in or around the middle of the eighteenth century (1750) in Pest, and made the first meerschaum pipe from a piece taken from Turkey by one of the ancestors of the present minister of foreign affairs for Austria, thus becoming the inventor of an industry that later spread on a very large scale. The first meerschaum pipe made by Kovács is kept in the Hungarian National Museum.[15]

This story is supported by another account, quoted in part:

> In the twenties of the eighteenth century, Count Andrássy brought back with him a wonderful piece of white mineral from his trip to the Balkans. The stone was light, floated in water and was so soft that one could carve it with a knife. The Count showed it to Kovács, a man known for his ability as a carver, and asked him to carve something for him...He thought that the material was ideal for making a pipe...Andrássy's pipe became famous in the elegant society of Budapest and Vienna...Kovács became snowed under with orders from aristocrats. In this way the adroit cobbler became the father of the world-famous pipe-carving art of Budapest and Vienna.[16]

But, according to Levárdy, this Budapest museum has neither this pipe nor a record of the donation![17] Furthermore, the only Hungarian count on record named Andrássy was born in 1823, exactly 100 years after the supposed debut of the first meerschaum pipe, yet the Andrássy-Kovács connection remained uncontested and unchallenged until the mid-point in this century. The first serious treatise on meerschaum manufacture was written by J. A. Tomas in 1799; the English-equivalent literal—and lengthy—translation of the title is *Practical Instruction to Manufacture Meerschaum Pipe Bowls, to Distinguish the Genuine From Artificial, About The Advantages to Boil Same in Wax and Tallow, to Break Them in and Also to Give Inferior Mass the Smoked Color.*[18] Tomas's research evidences that he had more than a passing understanding of the properties of meerschaum and the tools of the carver's trade a quarter-century before the storied cobbler allegedly produced the first meerschaum. If both the manuscript and the date of publication are authentic, then, indeed, a strong possibility exists that meerschaum pipe bowls were produced in that century, and Kovács could have been the very first individual in recorded history to have made two pipe bowls!

However, in 1949, E. Reid Duncan subverted the Kovács legend when he wrote a story in *Pipe Lovers.* He stated that "the greatest mystery of meerschaum pipes is the year in which this substance was first used in pipe-making."[19] He then claimed that in 1652 the French sculptor, Louis Pierre Puget, a student of the Italian sculptor, Berini, carved a large-scale meerschaum pipe titled *"Le Brigand au repos,"* (The Bandit at Rest). The pipe consisted of seven high-relief-carved, to-scale figures—a man, a woman, two children, two dogs, and an eagle. Duncan contended that he knew of no earlier account of a meerschaum pipe, but if this story is chronologically accurate, then the possibility also exists that if Puget carved this pipe, he may have modeled it after the work of earlier carvers, suggesting that meerschaum-carving was a 17th century, rather than an early 18th century, tradecraft. Duncan claimed that this pipe was made "...170 years before Count Andrássy was born."[20] If true, and if my math has not failed me, the only Count Andrássy of record (born in 1823) could not have given Kovács those two lumps of raw meerschaum at any time in the 18th century. And if Kovács was the individual to whom we collectors owe a debt of gratitude, perhaps an unnamed individual gave him those two samples.

Lap-style pipe bowl, bas-relief-carved scene of a deer hunt, reticulated chased silver windcap with finial, thumblift, and shank collar, 7" h., prob. German, ca. 1840. *Courtesy of the GC Collection.*

But let me challenge this thesis, because the story gets more complicated. A Neapolitan sculptor, architect, and painter, Giovanni Lorenzo Bernini (1598-1680), not Berini, may have inspired and influenced Puget, but Bernini was neither Puget's teacher nor mentor. The two never met. Puget (1622-1694), the son of a master mason, was not only a talented sculptor, but also a painter, decorator, military engineer, and ship designer. He did not serve under Bernini; he was under the tutelage of Pietro Da Cortona, a distinguished decorator of Italian palazzos. Puget's biographical sketch suggests that he had worked in stone and marble, not meerschaum; he was famous for buildings and statues, not tobacco pipes. And to this day, the rightful carver of le Brigand is not known, but

Lap-style pipe bowl, bas-relief-carved scene of the siege of a fortress, 5.25" h., en-suite meerschaum windcap with four cannon, prob. Austrian Empire, ca. 1850. *Courtesy of the GC Collection.*

I personally doubt that this pipe, given its extraordinary and crisp detail, finesse, and figural balance and proportionality, was carved at any time in the 17th century![21]

Many other intellectual excursions are possible, of course, but suffice to say that the precise timetable for meerschaum's first use is, at best, rather fuzzy, and I accept that the origin still remains enshrouded in myth, mystery, and make-believe. "When, where, and how meerschaum first came to be used as a material for making [pipe] bowls, are questions upon which history is silent. The name of its first discoverer is lost, and will never be known."[22] In a word, as this book is wont to emphasize, much is known, and not a little is unknown. I believe it is far more important to recognize that whatever circumstances led to its use and eventual popularity as a medium for pipes, meerschaum gradually asserted an incomparable superiority to any other material, such as clay, wood, and porcelain in use during the same period.

For further reading about the manufacture of meerschaum and amber, research points to eight early German manuscripts.[23] Affiliated with the pipe industry was a contingent of skilled wood turners, silversmiths, goldsmiths who manufactured findings, such as windcaps, retaining chains, and shank bands—or, variously, mounts or ferrules—for pipes and holders. Those interested in searching for the identity of the various smiths who worked hand-in-glove with the meerschaum carver will be best served by a book on hallmarks.

Lap-style pipe bowl, bas-relief-carved scene of the implements of war, and circumscribing these weapons are grotesque-looking faces of men, perhaps suggesting the futility of war? Bowl is terminated in a silver windcap with finial and collar, 4.5" h., prob. German, ca. 1860. *Courtesy of the GC Collection.*

Chapter IV. A Perspective on the Trade

The very earliest beginnings of the meerschaum trade are relatively clear, but a complete and accurate account of the trade's expansion, over time, cannot be depicted without some degree of conjecture and a few assumptions. I have always believed that this trade, like so many others of that era, probably maintained some business records for internal use—raw materials, sources of supply, wages, clients, sales, employee rolls, maybe a catalog of drawings and sketches, etc.—but in my search, I have found none of these. I suspect that the business records of most of these small factories are lost forever, and if such records exist today, they are accessible to very few people; I am not one of these fortunates. So, what I offer, then, are some insights that allow the reader to get a sensing of how this loose federation of pipe-makers was seen by others, and I leave the rest to imagination. These insights, as I mentioned earlier, are in the form of relevant and unaltered anecdotal quotations; I have taken no liberties with or embellished the information I have found in these few eye-witness reports.

In historical brevity, soon after the Seven-Year War (1756-1763), Ruhla was considered the first city in the Holy Roman Empire where meerschaum pipe bowls were manufactured; around 1800, Ruhla was a meerschaum trade center with an estimated 150 carvers working in 27 factories making bowls that were handed off to other tradesmen who supplied stems and mouthpieces. Much before the end of that century, Lemgo, another city of the Empire, established a small meerschaum pipe center, followed by a third city, Nürnberg. Almost concurrently, the idea of using meerschaum residue was conceived, and the production of lesser-grade imitation meerschaum bowls began.

In a late 19th century English view of the times, "the carving of those (meerschaum) blocks is executed entirely by foreigners; and even the pipes made in this country are given to either German or French artists to manipulate, which does not speak much for our native ability in this department of commerce."[24] "At New York both the manufacturers of meerschaums and their workmen are natives of Germany."[25] "Constantinople is the great mart for the sale of meerschaum, as Vienna is for its manufacture into pipes."[26] In the words of another: "The heydey of the Meerschaum pipe was in Vienna from around 1860 to the turn of the century, and it became more popular than the English clay pipe and the porcelain pipe of central Europe. It is likely that up to 80 percent of the finest Meerschaum carvings were done in that city which became the art's centre."[27] Wherever the search leads, the universal view was that Europe and, more specifically Vienna, was the hub of the meerschaum trade where ornamentation was a distinct branch of business:

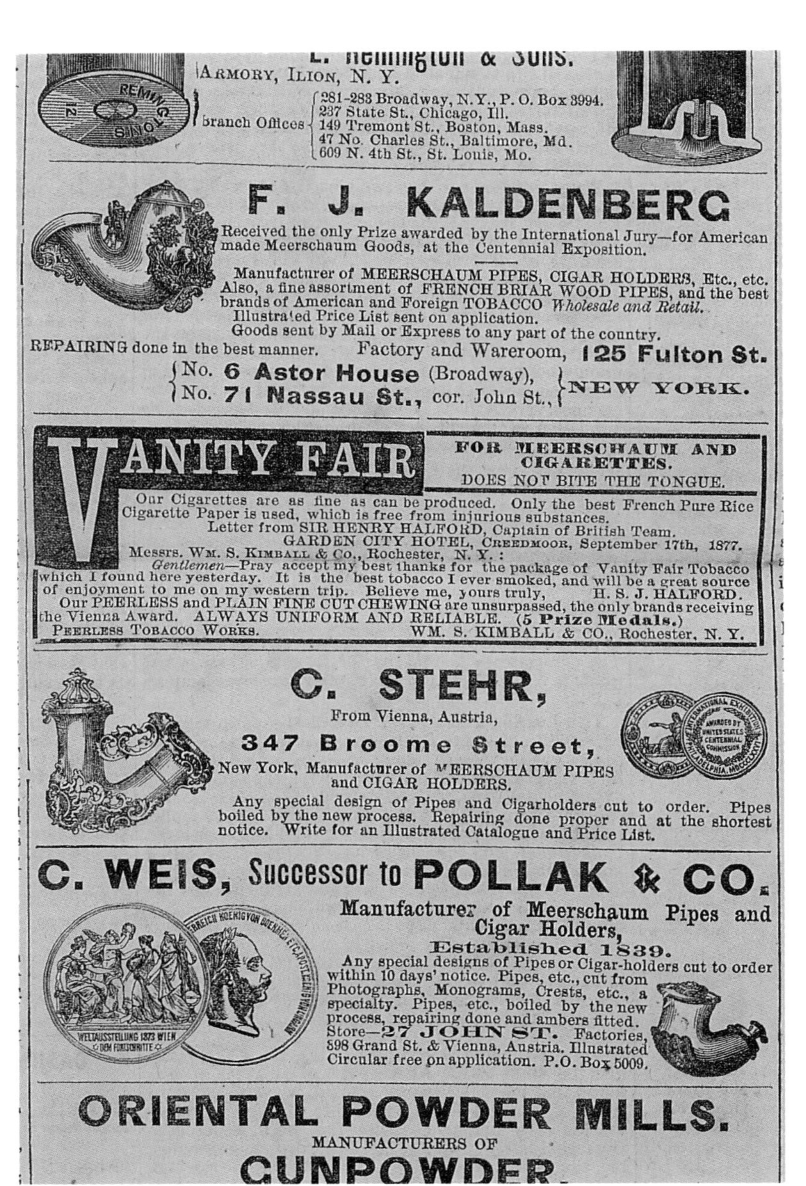

L. Remington & Sons.
ARMORY, ILION, N. Y.

Branch Offices:
- 281-283 Broadway, N.Y., P. O. Box 3994.
- 237 State St., Chicago, Ill.
- 149 Tremont St., Boston, Mass.
- 47 No. Charles St., Baltimore, Md.
- 609 N. 4th St., St. Louis, Mo.

F. J. KALDENBERG

Received the only Prize awarded by the International Jury—for American made Meerschaum Goods, at the Centennial Exposition.

Manufacturer of MEERSCHAUM PIPES, CIGAR HOLDERS, Etc., etc. Also, a fine assortment of FRENCH BRIAR WOOD PIPES, and the best brands of American and Foreign TOBACCO *Wholesale and Retail.*
Illustrated Price List sent on application.
Goods sent by Mail or Express to any part of the country.

REPAIRING done in the best manner. Factory and Wareroom, **125 Fulton St.**
No. **6 Astor House** (Broadway),
No. **71 Nassau St.,** cor. John St., NEW YORK.

VANITY FAIR

FOR MEERSCHAUM AND CIGARETTES.
DOES NOT BITE THE TONGUE.

Our Cigarettes are as fine as can be produced. Only the best French Pure Rice Cigarette Paper is used, which is free from injurious substances.
Letter from SIR HENRY HALFORD, Captain of British Team.
GARDEN CITY HOTEL, CREEDMOOR, September 17th, 1877.
Messrs. WM. S. KIMBALL & Co., Rochester, N. Y.:
Gentlemen—Pray accept my best thanks for the package of Vanity Fair Tobacco which I found here yesterday. It is the best tobacco I ever smoked, and will be a great source of enjoyment to me on my western trip. Believe me, yours truly, H. S. J. HALFORD.
Our PEERLESS and PLAIN FINE CUT CHEWING are unsurpassed, the only brands receiving the Vienna Award. ALWAYS UNIFORM AND RELIABLE. (5 **Prize Medals.**)
PEERLESS TOBACCO WORKS. WM. S. KIMBALL & CO., Rochester, N. Y.

C. STEHR,

From Vienna, Austria,
347 Broome Street,
New York, Manufacturer of MEERSCHAUM PIPES and CIGAR HOLDERS.

Any special design of Pipes and Cigarholders cut to order. Pipes boiled by the new process. Repairing done proper and at the shortest notice. Write for an Illustrated Catalogue and Price List.

C. WEIS, Successor to POLLAK & CO.

Manufacturer of Meerschaum Pipes and Cigar Holders,
Established 1839.

Any special designs of Pipes or Cigar-holders cut to order within 10 days' notice. Pipes, etc., cut from Photographs, Monograms, Crests, etc., a specialty. Pipes, etc., boiled by the new process, repairing done and ambers fitted. Store—**27 JOHN ST.** Factories, 398 Grand St. & Vienna, Austria. Illustrated Circular free on application. P.O. Box 5009.

ORIENTAL POWDER MILLS.

MANUFACTURERS OF
GUNPOWDER.

Advertisement from the newspaper *Army-Navy Journal. Gazette of the Regular and Volunteer Forces*, New York, February 2, 1878, 416. *Courtesy of S. Paul Jung.*

Cover of the Kaldenberg & Son Company, New York, catalog, 1868. *Courtesy of S. Paul Jung.*

The decoration of meerschaum is an art, and employs hundreds of workmen. Vienna is the headquarters of the meerschaum manufacture, and the Germans and Austrians have it almost entirely in their hands. Attracted by the higher wages offered in England, there are many Germans engaged in the manufacture of meerschaums in London. They frequently earn from £4 to £6 a week.[28]

Soon, talented European and American craftsmen alike set up shop and changed the complexion in smoking pipes. According to Raufer, the largest supplies of raw meerschaum blocks went to Vienna, Leipzig, Paris, and eventually, to the United States, although he emphasized that Vienna *"der Centralpunkt der Fabrikation ist und bleiben wird; ebenso war Wien die Wiege, von welcher der Meerschaum als Modeartikel feinen Ausgang nahm."*[29]

Until about 1850, meerschaum was practically unknown in the United States, but its popularity in Europe spread across the Atlantic and, soon thereafter, American interest in this pipe material caught on.

The manufacturing meerschaum industry was introduced in 1855, when the old business was a mystery, and no one, at that time, knew as much about it, as the reader can glean from this article. At that period it was almost impossible to obtain the raw material in this country. The late F. W. Kaldenberg was a mechanician, an all-around artisan and general workman of the kind that we do not meet with today, and he was approached by an Armenian, one Bedrossian, who had brought two cases of raw meerschaum into this country from Asia Minor. It was just as difficult for the Turk to find a purchaser for his material as it was for the artisan to find the material to make the pipes. Consequently the meeting of these two was a happy one, and furnished the missing link in the circle. It was not long before these two cases of meerschaum were turned into pipes of special shape and design, which brought the literati, the artistic, and the mercantile nabobs of the great City of New York, to the workshop of the artisan who had wrought the first meerschaum into pipes in the United States.[30]

Most other information about the evolution and expansion of the American meerschaum industry is scant, vague, and illusory. Hence, I have relied on the accounts of those closest to the trade, the relatives of a few famous carvers. After the publication of *A Complete Guide to Collecting Antique Pipes*, I initiated correspondence with the next of kin of four noted carving families, and I believe that

Advertisement from the G. Weiss Company, Trieste, Italy, ca. 1880. *Courtesy of the Author's Collection.*

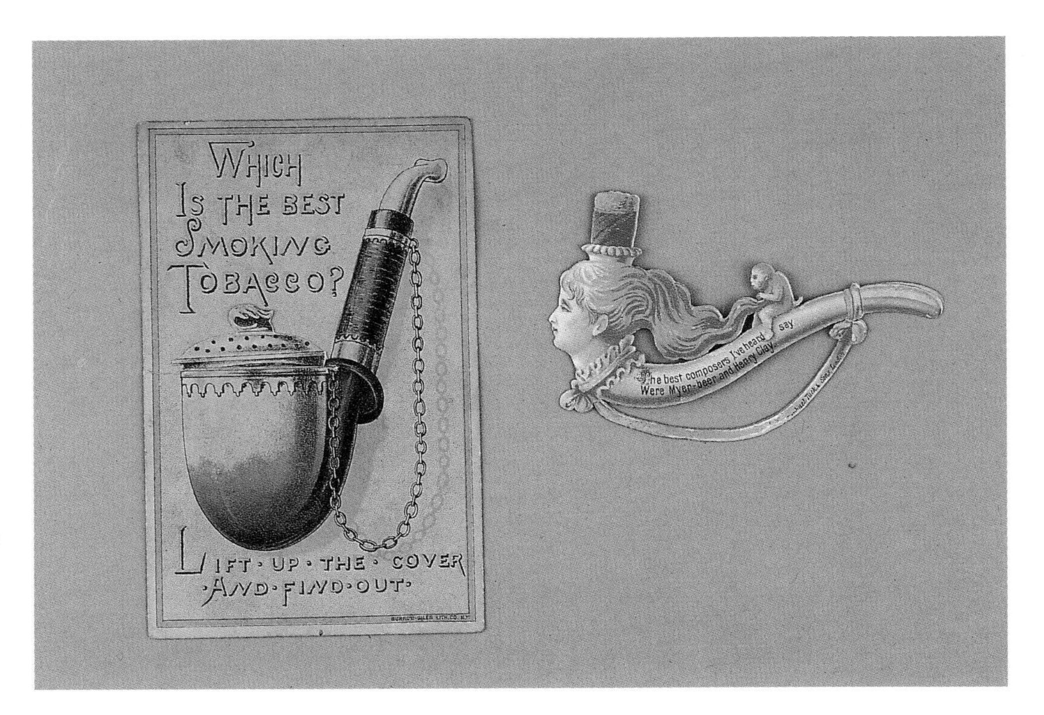

Industry trade cards. On the left, late 19th century advertisement for Blackwell's Genuine Durham tobacco; on the right, a die-cut of a cheroot holder designed by Raphael Tuck & Sons, London, England, ca. 1900. *Courtesy of the Author's Collection.*

the new information I have garnered offers a slightly better visualization of the times. In the earlier book, I mentioned three prominent American families, the Fischers of New York, the Fischers of Massachusetts, and William Demuth of New York City. The relatives of both Fischers were able to offer a few additional insights, but the descendants of William Demuth who now live in Washington, D.C., and in New York State had no recollection about the man or the business. I also offer a brief account on a family of French pipe-carvers, provided by Gilbert Guyot, the last of several generations to operate a retail tobacco shop in Paris.

William Demuth (1835-1911), a native of Germany, entered the United States at the age of 16 as a penniless immigrant and, after a series of odd jobs, found work as a clerk in the import business of a certain Edward Hen, a tobacco tradesman in New York City. In 1862, William established his own company in this city. The William Demuth Company specialized in pipes, smoker's requisites, cigar-store Indians, canes, and other carved objects. In 1887, William was joined by his half-brother, Leopold, who had emigrated from Darmstadt, Germany. At various times during the company's lengthy history, factories were located at different addresses throughout the city, including Richmond Hill (Queens), Glendale, and on Broadway and Fifth Avenue in Manhattan, where the retail showroom was located. By 1890, it was claimed that Demuth owned the largest pipe import, wholesale, and manufacturing business in the United States and, perhaps, the world. But, certainly, show figures and cigar-store Indians, many of which were carved by one of the employees, Samuel Anderson Robb, represented a significant fraction of the business. When William died, Leopold became president, and Louis, one of the three children of William and his wife, Harriet, oversaw the day-to-day operations until he died in 1912. By 1916, the factory had about 1,000 workers in its employ. The company was sold in 1925, and Leopold died 18 years later, in 1943.

Today, the Demuth Company is probably well known for the famous trademark, WDC, in an inverted equilateral triangle. But the meerschaum literati know a more important fact that William commissioned the figurative meerschaum Presidential series, 29 precision-carved likenesses of John Adams, the second president of the United States (1797-1801) to Herbert Hoover, the 30th president (1929-1933), and "Columbus Landing in America," a 32-inch-long centennial meerschaum masterpiece made in America that took two years to complete. The Presidential series was sparked by Demuth's friendship with President James A. Garfield, a

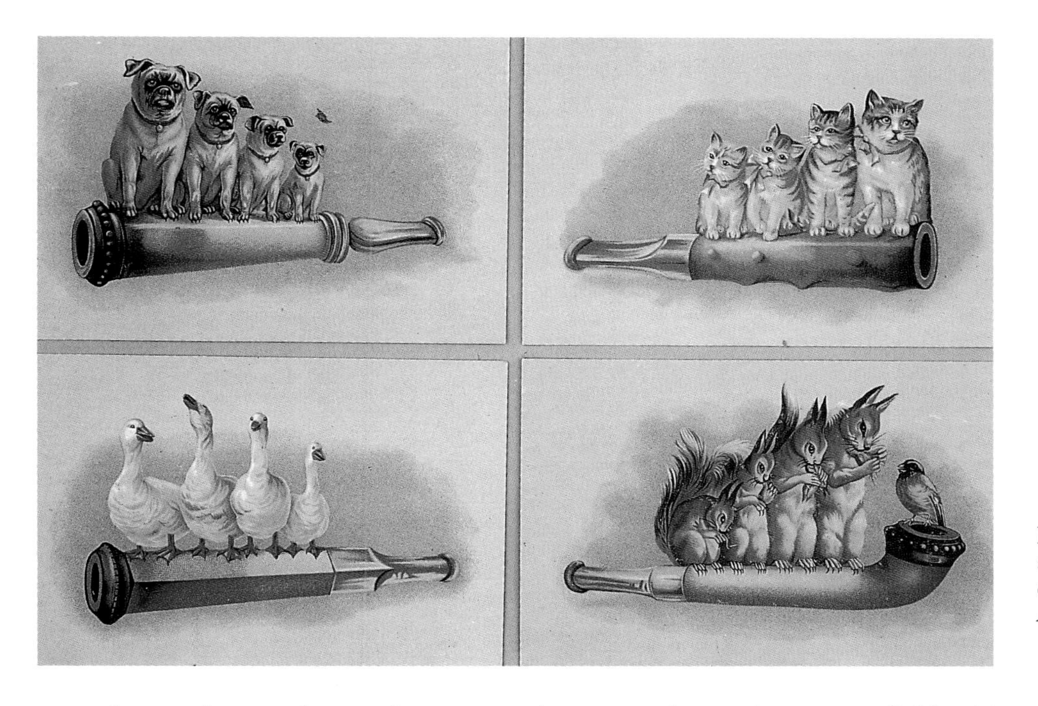

Four postcard advertisements for animal-adorned meerschaum pipes and cheroot holders. *Courtesy of the Author's Collection.*

connoisseur of meerschaum pipes; Demuth presented two pipes to Garfield at his inauguration in 1881, one in his likeness, the other in the likeness of the President's wife. Thereafter, Demuth arranged for another figurative precisely matching the others to be added to the collection as each new president acceded to the White House, terminating with President Hoover. A certain D. A. Schulte (Schulte Cigar Stores) purchased the company in 1925, and when the Demuth interests were later sold to the S. M. Frank & Company, the collection remained with Schulte. The American Tobacco Company purchased the collection in 1940 and renamed it the "Half and Half Collection" after its best-selling pipe tobacco that was introduced in 1926. In 1957, American Tobacco donated the "Half and Half Collection" to the Valentine Museum, the Museum of Life and History of (the City of) Richmond, Virginia. In 1992, after retaining it for 35 years, the Museum wrote the final chapter of the "Half and Half Collection" by dispersing it.

The Fischers of Orchard Park, New York, a family of pipe-carvers, owned the House of Fischer in that city for more than 40 years, according to Arthur C. Fischer, a sixth-generation pipe-maker, who passed away at his Venice, Florida, home in November 1997, just shy of his 93rd birthday. Arthur's great-great-great grandfather is said to have been called the John Hancock of European pipes, having been appointed in 1742 the official pipe-maker for the Prince of Saxony, a pipe lover, a tradition carried on by the next two generations in Europe. August G., of the fourth generation, and his two sons, Gustave A. and Otto, came to the United States after the Civil War, in 1867, and, soon thereafter, all three continued the craft in the employ of the William Demuth Company. In 1892, the family relocated to Buffalo and operated a small pipe shop for a time at 301 Main Street and later at 64 Broadway. Some years later, the House of Fischer was relocated to various addresses in Orchard Park, the last of which was 6226 Boston Ridge Road. The family retains a letter from President William McKinley acknowledging receipt of a meerschaum pipe, a figural bust of his likeness that Gustave A. made for the president's birthday in 1900. August G. also made several impressive masterpieces for the Pan American Exposition of 1901, a few of which appear in this book.[31] By 1934, the Fischer establishment, then at 15 West Eagle Street, was the only shop between New York and Cleveland still working with meerschaum and amber.[32] When Arthur sold the Orchard Park shop in 1979, ending a 235-year tradition, he dispersed most of the fine-quality meerschaums made by his father and grandfather.

Ironically, another family of pipe-carvers whose craftsmanship has always been much admired was also named Fischer, but this family was based in the area around Boston, Massachusetts, and was unrelated to the Fischers of Orchard Park. Gustav Fischer Sr. (1846-1937), entered the Vienna Academy of Design at the age of 11 to study sculpture. He apprenticed at painting, wood carving, and sculpture in stone, clay, and plaster of Paris, and began carving pipes at the age of 14. He came to the United States in 1880, after his skilled workmanship received acclaim when a group of meerschaum figurative Indians he was commissioned to make was exhibited at the Philadelphia Exposition in 1876. After a short stay in New York, in 1893, he moved his family to Boston where he continued to carve until his death at the age of 91. Coincidentally, while Gustave A. Fischer of Orchard Park was crafting the President McKinley pipe, Gustav Fischer Sr. of Boston was making one in the likeness of Admiral George Dewey who was visiting this city after the Spanish-American War. For a time, Gustav Sr. owned a shop on Tremont Street, then one at 199 Massachusetts Avenue, and later at 275 Massachusetts Avenue. However, he is best remembered as the "man in the window" when he began working at the 33 Court Street location of the David P. Ehrlich & Company pipe shop in the 1930s. He would be seated at a workbench carving meerschaums and briars to the delight of patrons and passers-by. Newspapers of the day claimed that he was the oldest active pipe-carver in the world, and one of the best who used only his imagination. In his lifetime he may have made several thousands of pipes, some taking only a day, others as much as a week. He was quoted as having said: "I never use a model.... My models come from my head." His "Battle of Bunker Hill," after Amos Doolittle's famous painting of the battle and the fall of Major General Joseph Warren, M.D., stands out as one very remarkable sculpture. It is a monumental two-pound, 34-inch-long meerschaum work of art with 31 high-relief-carved figures, three American flags, and one British flag, a pipe that he carved over a period of four years.[33] Gustav Fischer Jr. (1887-1975) began to apprentice with his father at 14, and then managed the store even after he retired in 1955 until his death at the age of 88. Today, many Fischer pipes are dispersed to private collections, the Bruce Museum, Greenwich, Connecticut and, until recently, the Museum of Tobacco Art and History, Nashville, Tennessee, but a significant number of the very largest and best, among them, the "Battle of Bunker Hill," are retained by family members.

An 1897advertisement for a smoker's set. *Courtesy of the Author's Collection.*

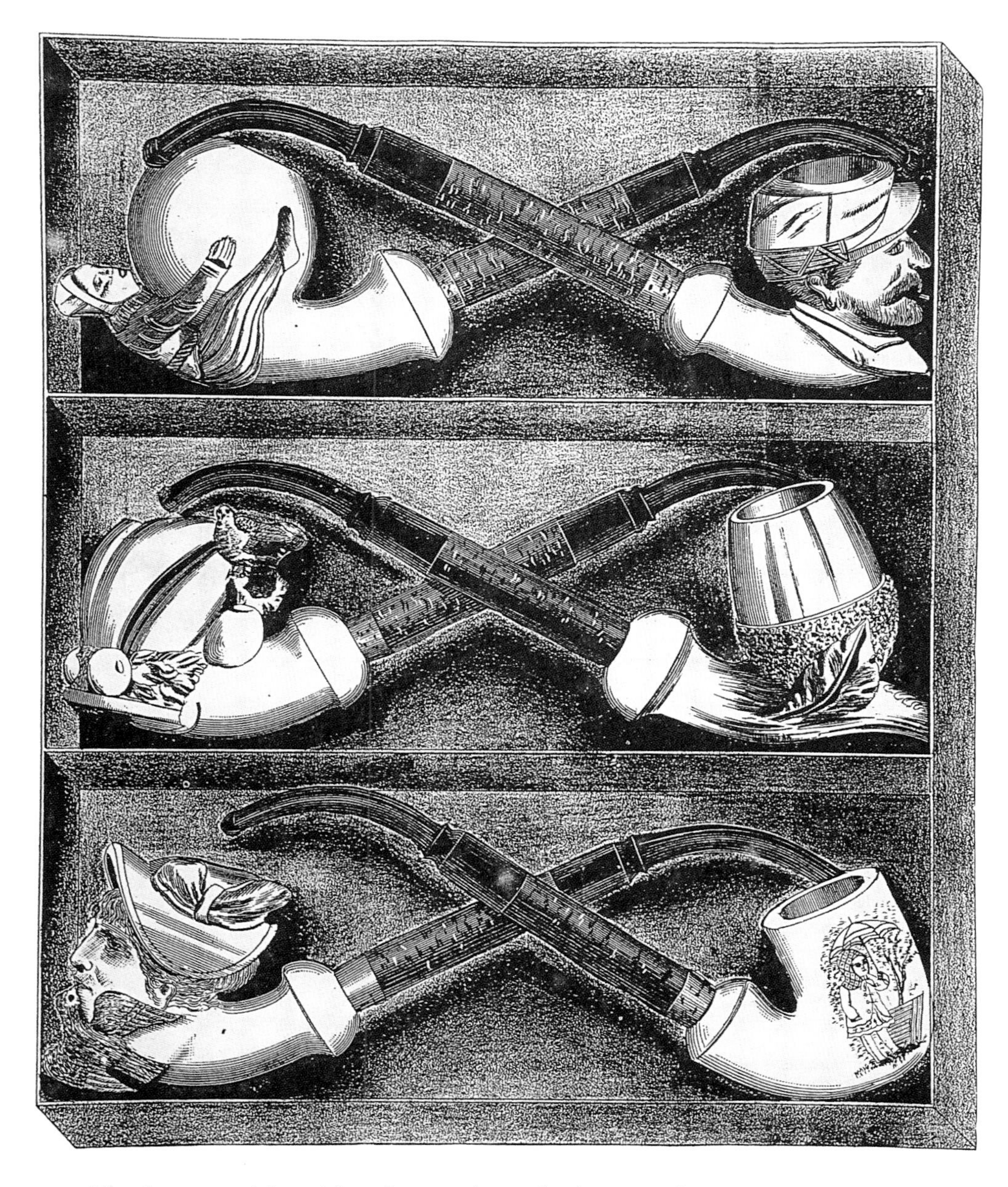

George Zorn & Co., *Pipes, Matches, Walking Canes, and Playing Cards. Illustrated Catalog*, ca. 1886-1887, 26. *Courtesy of S. Paul Jung.*

The last notable with whom I have had more than occasional contact is Gilbert Guyot, a fourth-generation pipe-carver, who closed his retail shop at 7 Avenue de Clichy in Paris in late 1995. His lineage dates to about 1825 when a certain Laporte sold his shop in the Palais Royal, Galerie de Chartres, Paris, to Marot Ainé. In 1851, Gilbert's grandfather, François (Frantz) Goltsche, an established master of meerschaum in Vienna, arrived in Paris and opened "l'Oriental," a shop in the same gallery around 1870. As best I understand the branches of the family tree from this point forward, François expanded retail operations with two more shops in the Paris environs, each managed by sons: Alexandre at "Aux 100,000 Pipes" on rue de Rivoli, and Auguste at "A la Sans Rivale," in the Passage des Panoramas. ("A l'Indien" in the Passage Jouffroy, Paris, may have also been affiliated in what might have been an early variant of the chain store.) Charles, one of Auguste's two sons, a specialist in amber and calabash pipes, succeeded François at the Palais Royal location, followed by Charles's brother, Roger, while Auguste opened another shop in 1906 at 7 Avenue de Clichy. Auguste married the sister of Henri Guyot, hence, the union of the Goltsches and the Guyots. Gilbert, Henri's son, assumed directorship in 1969, when his father retired, and continued on with the family tradition until 1995, but he only made and repaired briar pipes. Gilbert's son, Philippe, declined to follow in his

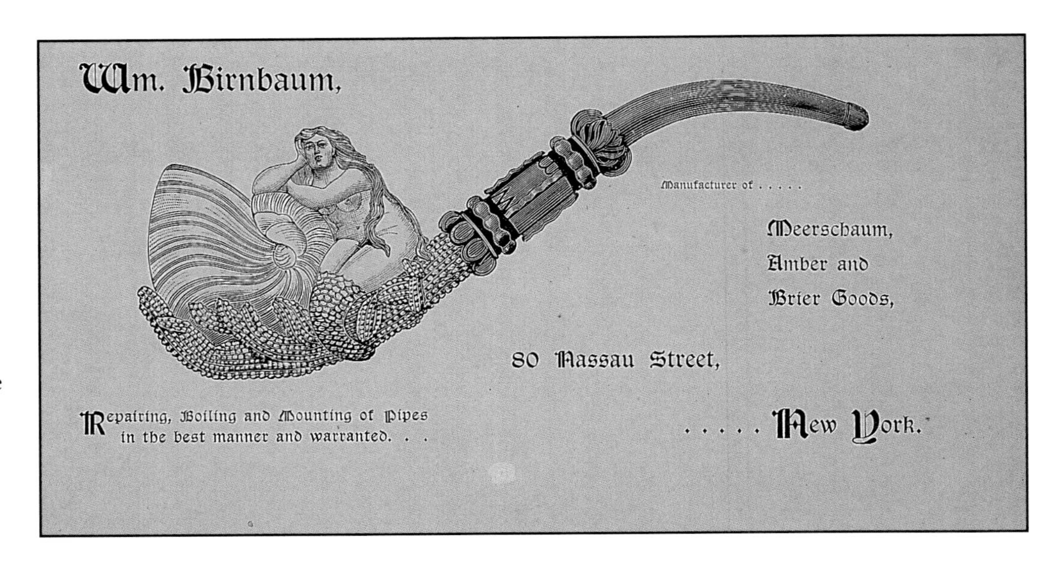

Advertisement for the William Birnbaum Company, New York, ca. 1900. The line drawing in the upper left-hand corner of the mermaid pipe is the company's trademark, and the pipe is, today, in a private collection. *Courtesy of the Author's Collection.*

father's footsteps, and so this almost 200-year heritage has ended. Gilbert's exceptional collection of antique meerschaums, principally figurative busts, that filled the store-front windows for years was sold privately. In one of his weak moments, he confided that, in his opinion, Sommer Frères, another Parisian firm of long-standing, produced the very best meerschaum pipes that France had ever offered the smoking public. I know that American collectors who search far and wide for Sommer Frères pipes would concur with Gilbert Guyot.

The brief exposition about the these four families is, in no way, meant to slight all the superlative European meerschaum carvers, such as Czapek, Hartmann, Schnally, and many others, or the handful of European immigrant-craftsmen adopted by America, such as Kaldenberg, Kutschera, and Roedel. Information about these notables is simply non-existent. So, for all those great carvers about whom we know nothing, let the pipes and holders attributed to these distinguished names bespeak their tradition and tradecraft.

Annex A is a compilation of names and locations of those affiliated with the meerschaum trade (although some may have only been distributors, wholesalers, retailers, and importers, and may have never once manipulated a pipe-carving tool). Even with a fitted case containing an embossed logo as proof, with so many unnamed individuals working the trade, I doubt that any piece can be ascribed to a specific individual. Hundreds of artists worked the trade, but "because pipemaking was an applied art, the name of even the finest creator or most talented technician was hardly ever affixed to an acknowledged masterpiece."[34] So view the list, which is by no means exhaustive, with some caution, and check the fitted cases of your meerschaums with the names that appear in this table.

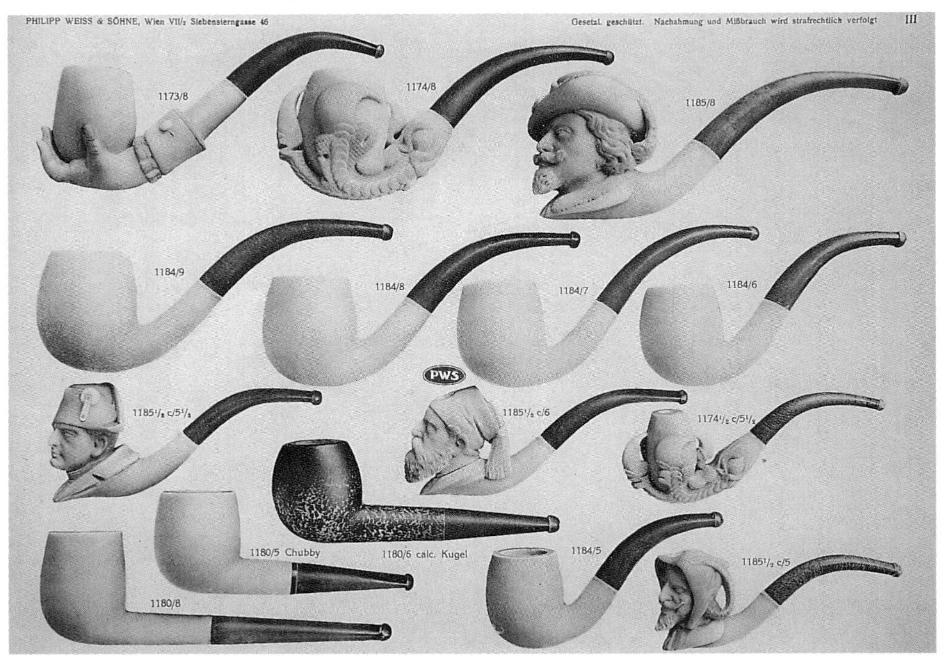

Philipp Weiss & Söhne, Wien-Berlin, pipe catalog, 1879, III. *Courtesy of the Author's Collection.*

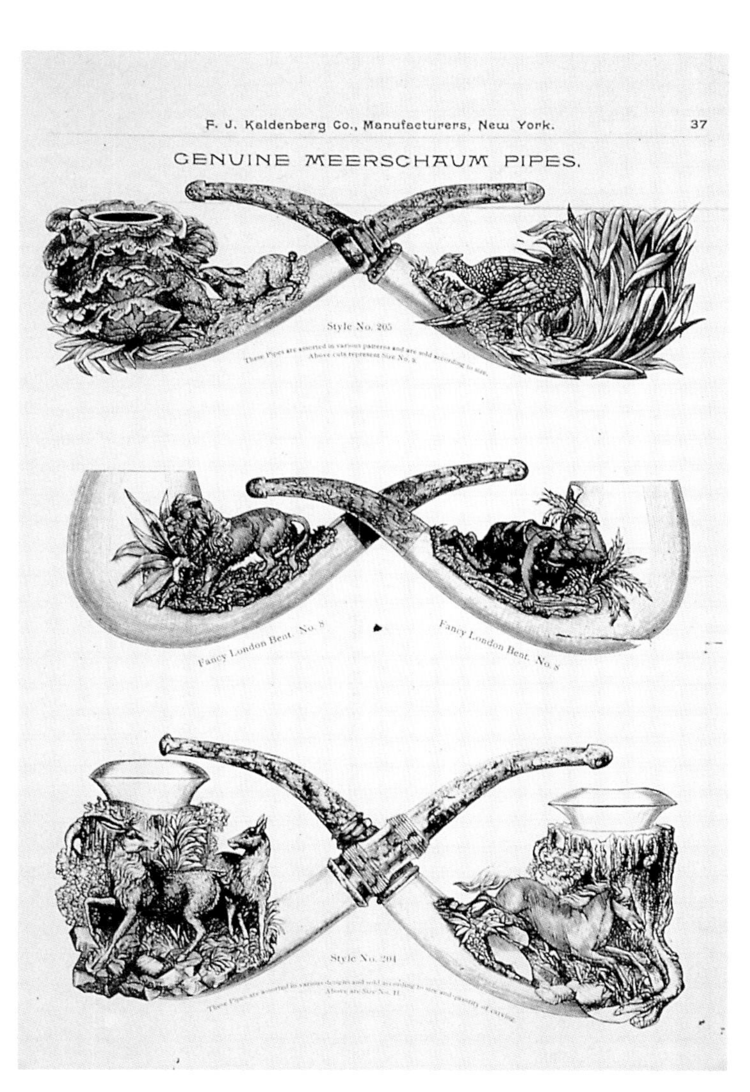

F. J. Kaldenberg Company, New York, catalog, ca. 1880, 37. *Courtesy of S. Paul Jung.*

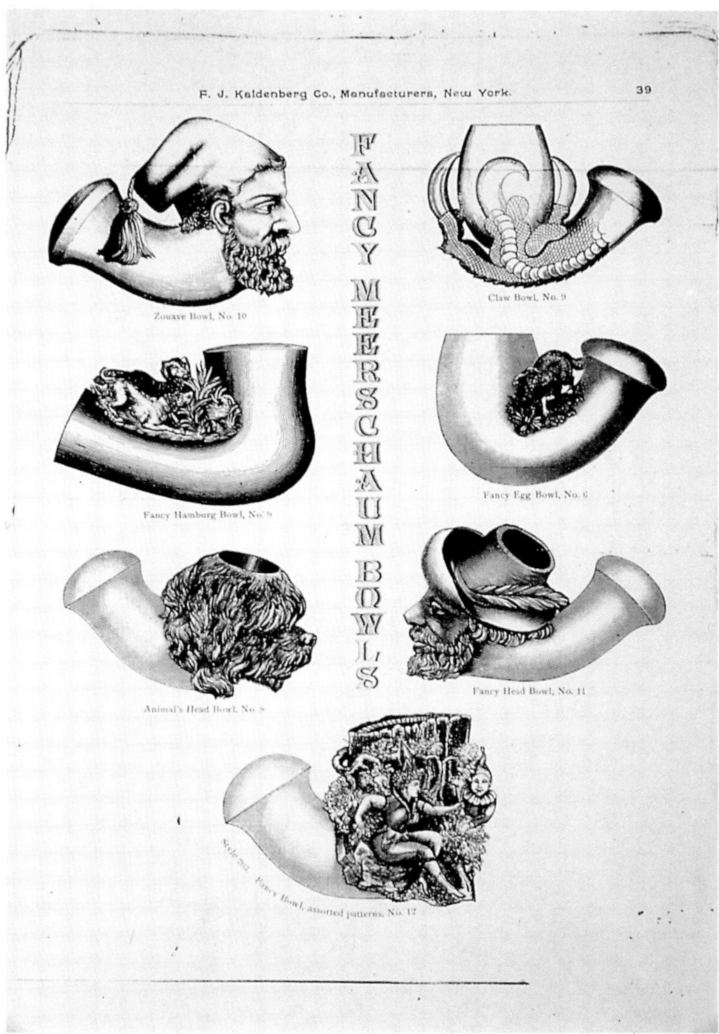

F. J. Kaldenberg Company, New York, catalog, ca. 1880, 39. *Courtesy of S. Paul Jung.*

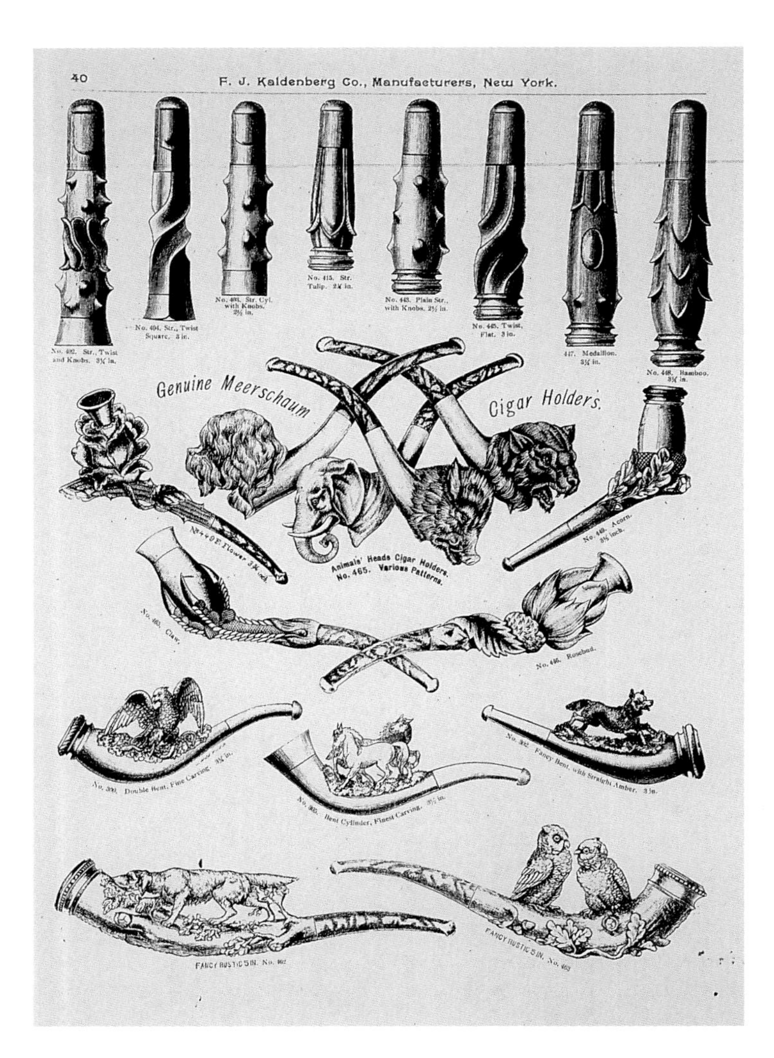

F. J. Kaldenberg Company, New York, catalog, ca. 1880, 40. *Courtesy of S. Paul Jung.*

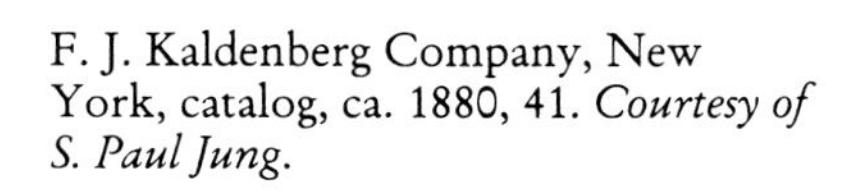

F. J. Kaldenberg Company, New York, catalog, ca. 1880, 41. *Courtesy of S. Paul Jung.*

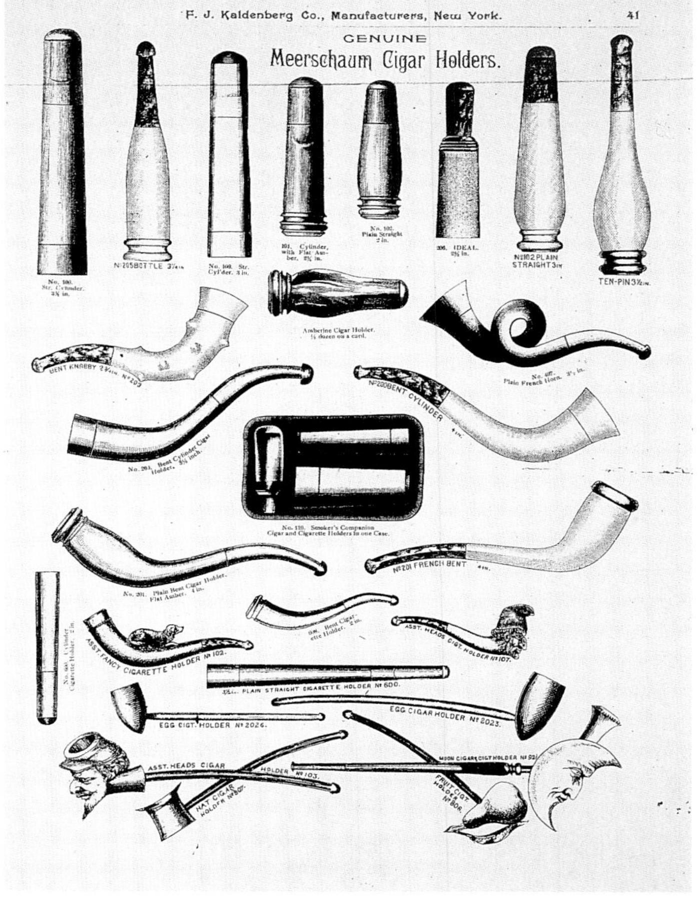

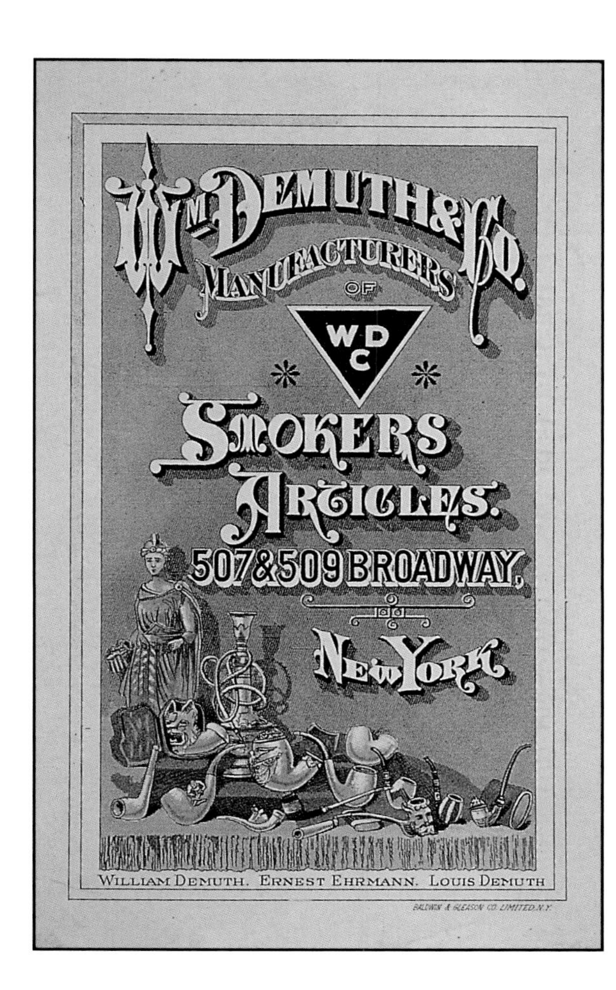

Advertising postcard, William Demuth Company, New York, ca. 1900. On the reverse side is an illustration of the pipe, Columbus Landing in America. *Courtesy of Tom Clasen.*

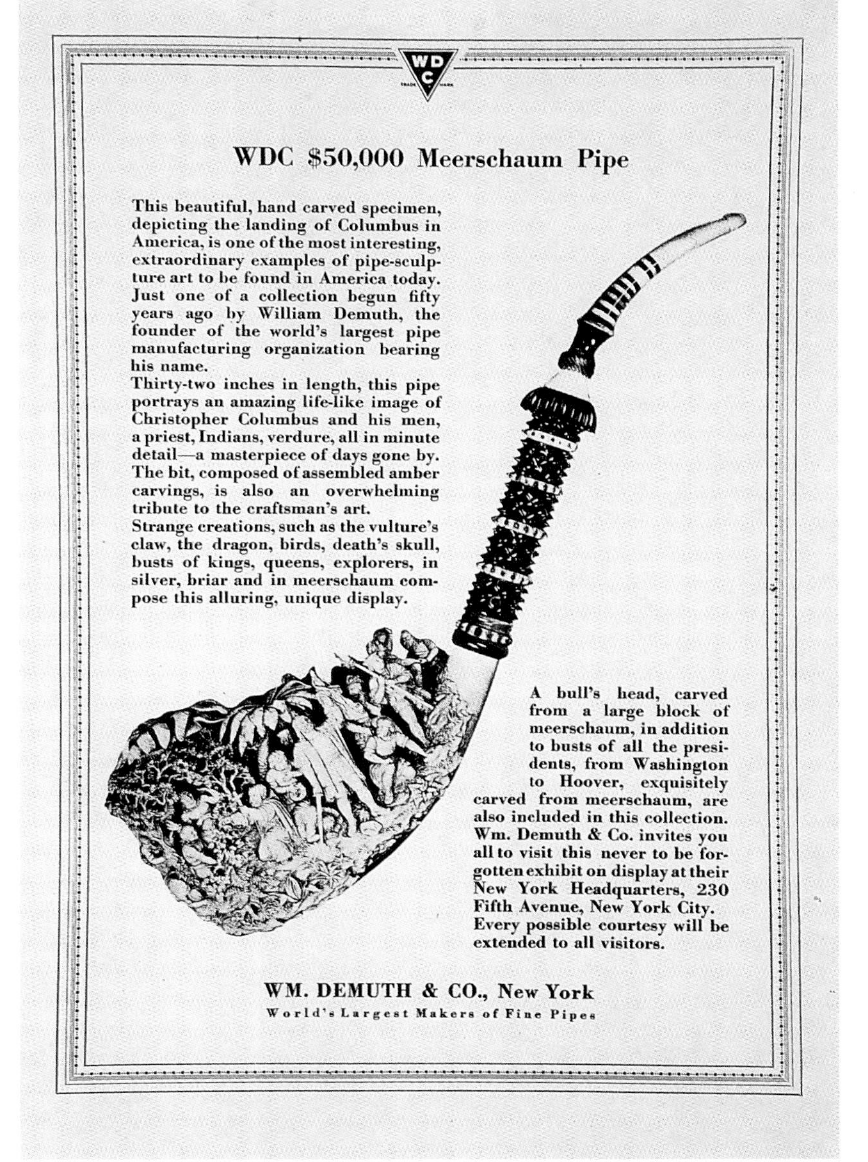

Page from the William Demuth & Company catalog, Spring 1932. *Courtesy of the Author's Collection.*

Portrait of Gustav Fischer, Sr., of Boston, Massachusetts, ca. 1930. *Courtesy of Thelma Fischer Jones.*

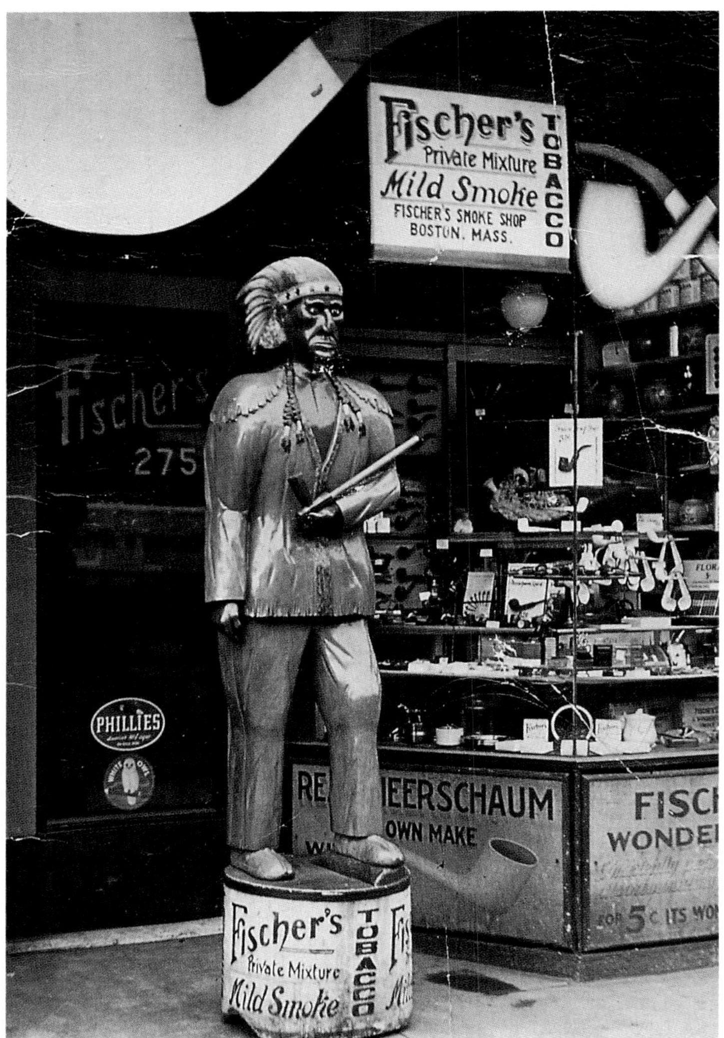

Storefront of Gustav Fischer's Smoke Shop, 275 Massachusetts Avenue, Boston, Massachusetts. *Courtesy of Thelma Fischer Jones.*

Chapter V. Dating and Attribution: The Metamorphosis of the Meerschaum

"Dating of meerschaum pipes is difficult and not particularly rewarding."[35] I do not completely agree with this assertion! Dating—and provenance and attribution—are not the exclusive craft of crystal-ball gazers and seers. Common sense and a degree of historic sensibility will serve quite nicely in identifying the approximate age and, perhaps, the origin of an antique meerschaum. Permit me to explain, although I may be on tenuous ground, and I am not confident that other knowledgeable collectors share my views. What I offer is logical and rational, and I am prepared to stand my ground on the thesis that the meerschaum pipe, from its introduction sometime around the early 18th century to the early 1900s experienced a three-stage metamorphosis, and each stage represented a revolutionary and dramatic change in form, shape, structure, character, and appearance. Each mutation was abrupt and permanent. When the cheroot holder and cigarette holder, mentioned later, were introduced in the latter half of the 19th century, their respective shapes were fundamentally influenced by the second stage of the pipe's metamorphosis. The final stage occurred around the beginning of the 20th century, when the briar pipe had achieved universal acceptance as the material of choice, and almost all its shapes were of simple construction. If this thesis is valid, then meerschaums can be approximately dated.

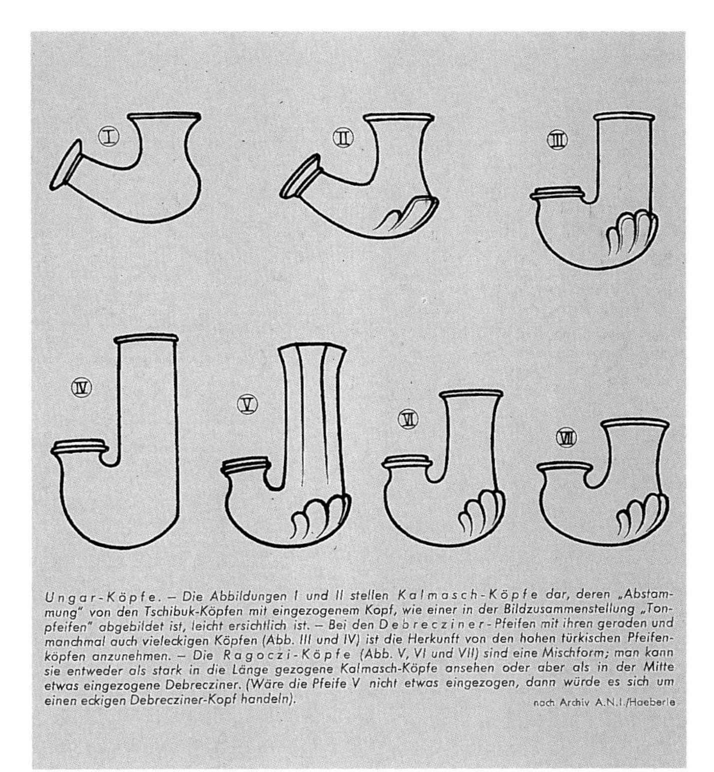

Ungar-Köpfe. – Die Abbildungen I und II stellen Kalmasch-Köpfe dar, deren „Abstammung" von den Tschibuk-Köpfen mit eingezogenem Kopf, wie einer in der Bildzusammenstellung „Tonpfeifen" abgebildet ist, leicht ersichtlich ist. – Bei den Debrecziner-Pfeifen mit ihren geraden und manchmal auch vieleckigen Köpfen (Abb. III und IV) ist die Herkunft von den hohen türkischen Pfeifenköpfen anzunehmen. – Die Ragoczi-Köpfe (Abb. V, VI und VII) sind eine Mischform; man kann sie entweder als stark in die Länge gezogene Kalmasch-Köpfe ansehen oder aber als in der Mitte etwas eingezogene Debrecziner. (Wäre die Pfeife V nicht etwas eingezogen, dann würde es sich um einen eckigen Debrecziner-Kopf handeln).

nach Archiv A.N.I./Haeberle

The three basic designs of Hungarian pipe bowls and their variants—Kalmasch (I and II), Debrecen (III and IV), and Rákcózi (V, VI, and VII). Helmuth Aschenbrenner, *Tabak von A bis Z*, 1966, 85.

Salmon & Gluckstein, *Illustrated Guide for Smokers*, ca. 1899, 85. *Courtesy of the Author's Collection.*

When I was just beginning to collect antique pipes, I was intimidated by someone who, I believed, had to be the most knowledgeable person I had met to that time. He assured me that he could identify the maker of any meerschaum he encountered. I was impressed! When I confronted him with one of my own, he could not provide a provenance because I did not have the fitted case. Aha! His secret pseudo-expertise was revealed. Using that person's standard of erudition, anyone able to read the embossed label in the fitted case would be an instant expert. The fitted cases of certain manufactories also revealed whether they exhibited competitively, because the proud recipients of medals and prizes at these exhibits embossed this information on the satin or plush lining inside the case. Often, the fitted case confounds, rather than clarifies, an antique meerschaum's origin. For example, the preface of a late 19th century trade catalog instructed the buyer that: "Pipes and cases can be stamped according to the wishes of customers with initials and any other marks. It will be sufficient to send a small sketch of the style of stamp desired with wording or letters."[36] Quite often, the fitted case contained no distinguishing trademark or logo inside or outside and, as often, the case has long since disappeared.

Meerschaums of that era—as opposed to today's Turkish pipes—were almost never signed. So, how does one know where the pipe was made, approximately when, and by whom? An exact method to determine the derivation of antique meerschaums has not been devised, and if one were posited, it would be, in my opinion, unsubstantiated and specious. Why? It is almost impossible to determine the provenance of a pipe or holder with even the slightest degree of conviction and certitude without the corroborating evidence of an original bill of sale, a carver's proof drawing, or the best—and most elusive—proof, an illustrated manufacturer's or exporter's catalog.

My approach to dating, provenance, and attribution of these antiquities is logical, but not supportable by concrete evidence or proof. I have studied enough meerschaums, however, to posit an approximate timetable based on what I believe was the stylistic metamorphosis, or transformation, of the meerschaum's size and shape over time. My belief is that just about every meerschaum pipe bowl made in the late 18th century to about 1850—give or take a decade on either side—was Hungarian-style, a near-cylindrical, plain or bas-relief-carved pot-shape. In fact, almost all pre-1850 meerschaum pipe bowls that exhibited carving were in bas-relief (low-relief). High-relief artistry, an art form in which the sculptured figures project from the background by half or more than half their full natural depth, was a movement that evolved around 1850, as craftsmen became more adept, more skilled, and more audacious at their tradecraft.

Precise names were attributed to the several assorted shapes of the bowl, such as Kalmasch (or Kalmas), Debrecen (a city in Hungary), and Rákcózi (or Rágóczy), named after a Hungarian prince. These three standard shapes, their variants, and their associated names were adopted by many other European carving centers. Collectively, these were described as having a "truncated pyramidical bowl, naturally with the stylistic, decorative modifications on the surface."[37] To make the pipe functionally complete, a long, tapered, wood, turned ivory, or turned horn stem with a horn mouthpiece, findings most often made by turners, was inserted into the back of the bowl. These two components were then paired with a retaining chain or cord to ensure that the bowl did not fall away from the stem and become damaged. Soon, names were devised to identify the assembled pipe, such as lap pipe or shepherd pipe. A few manufacturers incised the company name on the collar of the bowl; Adler and Weiss (or Weisz) of Budapest, and Koppa of

Rákcózi pipe bowl, circumscribed with accented scroll work, high-domed, chased silver reticulated windcap and silver collar band, 8" l., 7.5" h., ca. 1850. Smoking case is stamped Weiss. es Fla, Budapest. *Courtesy of the Author's Collection.*

Debrecen pipe bowl, bas-relief-carved portrait of Frederick the Great (Frederick II, King of Prussia, 1740-1786), on his mount, chased silver windcap, 6" h., and wood push stem, prob. German, ca. 1850. Carving attributed to Carl Wickenburg. *Courtesy of the GC Collection.*

Kalmasch pipe bowl, bas-relief-carved in-the-round battle scene led by the Iron Duke (British general and statesman, Arthur Wellesley, 1st Duke of Wellington), chased silver windcap, 5" l., 5" h., accompanied by a turned wood and horn push stem, fabric-covered horsehair flexible hose, and red amber duckbill mouthpiece, prob. Austrian Empire, ca. 1840. *Courtesy of the Author's Collection.*

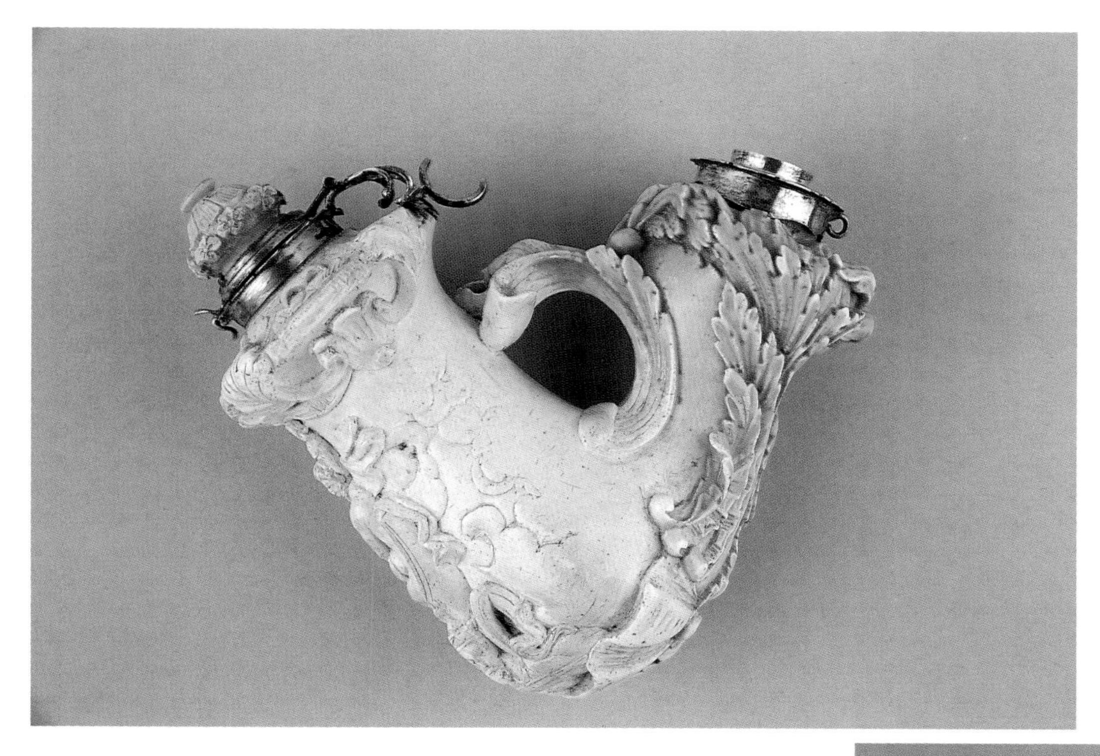

Debrecen pipe bowl, bas-relief in-the-round scene of Neptune, the Roman god of the sea (Greek god Poseidon) and Amphirite (a granddaughter of the Titon Oceanus), acanthus leaf accents on underside, silver windcap with meerschaum finial, silver shank collar, 4.5" l., 5" h., prob. Austrian Empire. Pipe is incised "Joseph Schwager, 1837." *Courtesy of the FB Collection.*

Dresden are three of the trademarks most often found. Occasionally, a fitted smoking case (to cover the body of the bowl, but not the tobacco "hole") accompanied the pipe, permitting the user to smoke without ever touching the surface of the bowl. The pipe bowls on pages 33 to 35 (top) are early to mid-19th century Hungarian-style lap pipe bowls.

Then came the pipe's transformation from the lap-style pipe bowl with push stem, to one of reduced proportions, an articulated bowl and mouthpiece. For about the next 75 years, a smaller, more portable pipe, executed with flair and extravagance, became the standard, and the long wood, horn, or ivory stem was replaced by a short amber mouthpiece, creating a perfect alliance of elegance and color contrast. This is the period in which the expressions in meerschaum represented, in my opinion, the acme of gifted European and American craftsmanship. A snug-fitting carrying case was made to the precise height and length of the pipe, and the carver (or the retailer, exporter/importer, distributor or wholesaler) embossed his logo or trademark on the plush or satin lining inside the case.[38] During this period, then, the attribution, at least from the better manufactories, was almost always embossed or stamped in the fitted case. The illustration on page 36 is a typical late 19th-

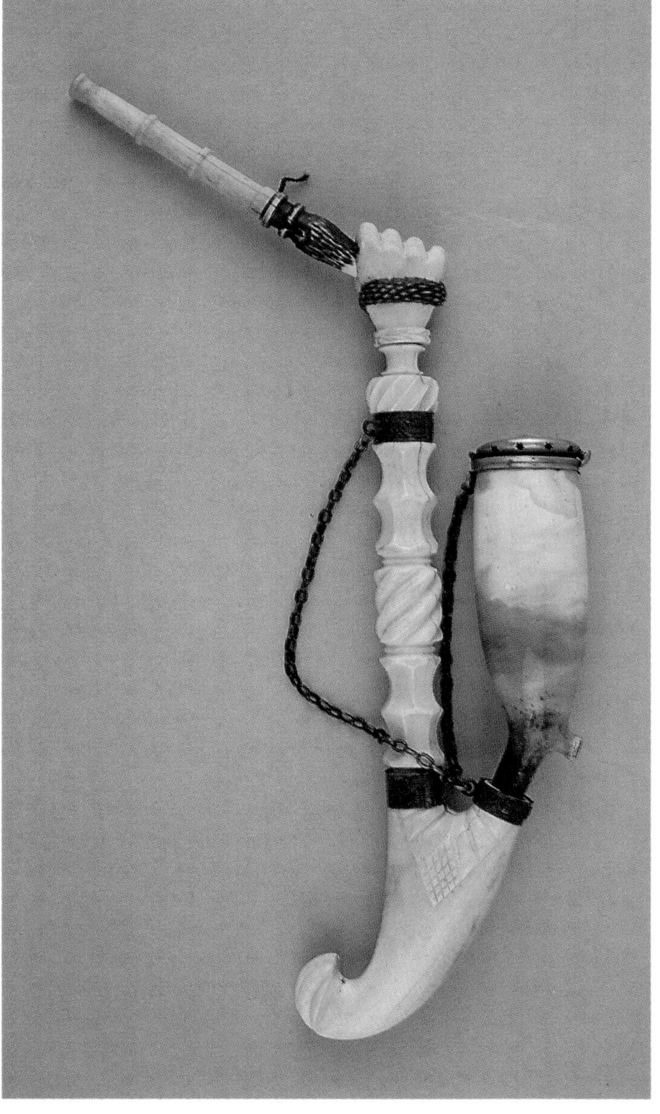

This style, most often manufactured in porcelain or wood, known as a *gesteckpfeife* (consisting of parts), could be assembled with any of five variant bowl shapes. This bowl is meerschaum, and the reservoir (beneath the bowl) and the stem are turned ivory, 14" h., German, ca. 1860. *Courtesy of the Author's Collection.*

This is a premier example of meerschaum craftsmanship at its acme. Note the exquisitely crafted anatomic art expression, the lifelike visage of this figurative head of Bacchus, the god of wine, displaying an advanced state of inebriety, amber accents on shank, 13.75" l., 5.5" h., prob. French, ca. 1890. *Courtesy of a WC Collector.*

century meerschaum pipe, and it is not difficult to see the transition from the shape and style in the previous four illustrations. During this time, everything imaginable was expressed in meerschaum at least once and copied many times over.

In this same time frame, tobacco was manifested in two other mediums, little cigars and cigarettes, and both influenced slight changes in the meerschaum trade, but neither resulted in a further mutation of the pipe's format. By 1850, cigars were already an accepted mode of smoking, and the popularity of the meerschaum pipe spun off into the cheroot holder, known as a smoking or cigar tube in England, and a segar holder or cigar-pipe in various other locales. The cheroot was a small cigar characterized by its straight body and completely open ends. It was normally round in cross-section, although a small square-pressed cheroot was also manufactured. The cheroot became a popular smoke of the period, and the most popular holder designed for it was of meerschaum, although other materials such as amber, bone, ivory, tortoiseshell, and wood were also used. The holder was simply a diminutively-sized pipe, as shown on the opposite page. Not unexpectedly, these, too, were executed with flair and fantasy, and were offered in great variety. As Fairholt wrote:

> This (the cigar) has led to the invention of cigar-tubes, or holders, by which they are kept at a distance from the lips, and may be entirely consumed. These are sometimes made of meerschaum clay, and occasionally decorated with sculptured figures, as in the example engraved, where a lion is baited by dogs. Such sculpture serves another purpose than mere ornament, as it gives the smoker's fingers a firmer hold, for which reason the scroll of foliage beneath has also been introduced.[39]

In a recently published book on cigars, the author, a cigar aficionado, had this to say:

> ...craftsmen allowed their fantasy to run riot. To judge from the examples that have survived this may have been as much for competitive display in the smoking room as for any other reason. In the most elaborate examples, the function of the piece may be almost masked by the sculptural adornment. It would seem that holders were generally used for the smaller sizes of cigars and, because variants of the tapering "torpedo" shape were favoured in the last century, the collector will notice that the cup, hollowed at the end of the "tube" to receive the cigar, is shaped accordingly.[40]

Those who have written about the social and cultural history of smoking generally agree that the English and French soldiers who fought in the Crimean War (1853-1856) learned how to smoke fine-cut tobacco wrapped in paper—the cigarette—from the Turks and Russians. "Importers in Paris and London were the first to introduce the 'original cigarette' from Petersburg and Constantinople into Germany."[41] Again, the meerschaum worker adapted his skills, having learned to craft the small cheroot holder with diligence; he "downsized," so to speak, to accommodate a new tobacco product, making a small, straight smoking tube (See below) that could be used with the newly popularized cigarette.

In a short span of about 10 to15 years after both the cheroot and the cigarette were accepted in social circles, meerschaum was in very great demand, because cigarette, cigar, and pipe smoker alike sought this mineral as the material *du jour*. Thus, these 75 years were probably a period of heightened manufacturing activity and competition in the trade, each meerschaum carver offering a wide range of

These utensils are, respectively, a cigarette holder (top), and a cheroot holder (bottom). Their composition and size are smaller than that of a pipe, but the concept of high art is unchanged. The straight-stem cigarette holder portrays a young trainer coaxing a bear, 5" l., 2" h., Au Phenix, Marseilles, ca. 1885. Similarly, the cheroot holder depicts a circus clown and a bear holding a broom, 5.5" l., 2.5" h., prob. German, ca. 1880. The dark color of the removable meerschaum coloring bowl into which the cheroot was inserted (in this specimen, the bowl is located at the back of the bear), the metamorphosis of milky-white meerschaum to tobacco brown, is caused by the absorption of tobacco tar. *Courtesy of the Author's Collection.*

This is an example of the style of pipe made during the last stage of the golden age of meerschaum. If one were to remove the high-relief-carved medieval troubadour serenading a young lady, this pipe would essentially be a "plain vanilla" traditional shape that is, today, known as a billiard, 7.25" l., 3" h., Au Pacha le Nouvel, Paris, ca. 1910. *Courtesy of the SP Collection.*

motifs for three discrete smoker's accouterments, one for each type of tobacco consumer, each slightly different in shape and size, and each displaying some degree of decoration and adornment. As with the pipe, a fitted case accompanied the cheroot holder and the cigarette holder.

I believe that the last stage of the meerschaum's metamorphosis, or mutation, occurred around 1900, give or take a decade—the beginning of the end of the golden age of this craft—and this final change was prompted by two different influences. The first, and most obvious, influence was the surge in popularity of a competing pipe material, briar, the root of the heath tree, discovered as a practical medium for pipes in the mid-1850s. Interestingly, in the 150 years of briar pipe production and use since then, little has changed. Except for the extraordinary quality of today's craftsmanship and the emphasis on high-grade grain, the briars made now are not much different in style or shape than those made in the last century; the briar has always been an uncomplicated utensil with minimal, subdued, or no decor, and its shapes, in relation to the meerschaum of this period, were of traditional or conservative design. Thus, to be competitive, the meerschaum carver had to restrain his carving technique, repress his penchant for flair and fancy, and align the appearance of his product more closely to that of the briar, a new medium that was rapidly being accepted as the "best of class" for the pipe smoker.

The second influence for the third and last change is more subtle. I contend that the coalescence of cosmopolitan life, industrialized society, and urbanization dictated the third mutation of the meerschaum pipe and both holders. After all, no self-respecting post-Victorian smoker would want to be seen in public places smoking a meerschaum pipe with the carved image of Diana and the Hunt encircled by feral animals, unless he desired the attention of and, more likely, ridicule from other urbanites. Such pipes were now to be relegated to the drawing room, the library, the den. On the street, in the office, shop, and pub, the meerschaum smoker was obliged to use a more conservative-looking utensil. Hence, the meerschaum craftsman, once again, effected another adaptation for the times: the carved portion of the meerschaum pipe either retreated to the rear of the bowl astride the shank, or to the front of the bowl, and the carving on both holders gradually disappeared altogether. Notice that the pipe in the illustration above lacks significant bas- or high-relief ornamentation—and no festoons, scroll work, or curlicues—that typified pipes of the previous half-century. If this decor were removed entirely, the pipe would not differ very much, other than in material, from its briar counterpart. I am also of the opinion that almost all meerschaum pipes adorned with the often stilted or stylized decor situated at the intersection of the bowl and shank were probably manufactured just after 1900 and later.

What I postulate is not new, and I am not alone in this belief, but I must include the opinion of J. W. Cundall who, in 1901, contended that the change occurred because "meerschaum is no longer so fashionable as it once was. Probably the numerous cheap imitations, the extreme readiness to break of the genuine article, and the fantastic designs into which it was fashioned, have led to its decadence in public estimation."[42] Whatever kindled the movement away from ornately carved implements of smoke, the stage was set for the decline and fall of high-art meerschaums.

Poring through catalogs and analyzing the offerings of this century, I am convinced that, after the first quarter of this century, both the shape and style of the finely crafted meerschaum pipe mirrored that of the smooth, unadorned briar. The cheroot holder eventually became passé as larger and longer cigars became vogue. Hard rubber, composition, horn, celluloid, and other exotic-sounding materials such as Chinese amber and redolite (which eventually gave way to bakelite, Delrin, lucite, and other pheno-resins), replaced genuine amber as the mouthpiece. Fitted cases were no longer produced. The meerschaum industry declined gracefully...it did not die; Turkey later brought this art form back to life. A brief article about the meerschaum trade appeared in the *Illustrierte Kronen-Zeitung* (a Vienna, Austria tabloid) on May 31, 1935. It reported that Vienna had more than 50 master-carvers, each supported by 10 to 15 worker-apprentices, in the last half of the 19th century but, in 1935, only four or five active carvers remained in business.[43] In the 1920s to 1930s, a few manufacturers here and abroad were still producing both the early Hungarian-style lap pipes and small cheroot holders, but this revival did not last long. Some 30 years later, Turkey became and is now still the principal center of meerschaum pipe manufacture, and its exports are aimed at resuscitating pipe-smoker and pipe-collector interest with reproductions of ornately-carved implements of smoke. But in the opinion of most serious collectors of antique meerschaums, Turkish carving is not even a good imitation.

Provenance is slightly more suited to a rational, but rather subjective, discussion. Provenance simply means source, origin, or ownership. From reading, traveling, and visiting fellow collectors, I have concluded that more than the Atlantic Ocean separated America and Europe in the last century. Taste also separated the two continents, because the popular talon and egg, the cuffed hand with egg, the skull-in-hand, the spread-eagled female nude known as Victory, the fleeting deer and stag, the recumbent dog, the rampant horse, and similar motifs, all were American expressions. Many thousands must have been produced, because they are found everywhere today. Another typical American style was the large, tulip-shaped bowl with bas- or high-relief carving on the obverse, a style that I call bi-dimensional, because the reverse side of the pipe bowl is relatively unadorned, indicating that the process of carving was somewhat hurried, not unlike assembly-line or mass-produced pipes. Furthermore, other than for some very impressive exceptions from one or two American carvers, the number of motifs was limited to those previously mentioned and figuratives (heads and busts of man, woman, and beast), particularly our native American stalking assorted quarry, and a handful of other less-than-imaginative subjects. This is not to say that American-made meerschaums of this period were low-grade or of poor quality; stated succinctly, a very large percentage seem to echo a mere handful of redundant themes.

The evidence from my research shows that two factors prevailed in European-made pipes: very detailed and highly intricate carving and very imaginative subject matter. Except for figurative heads, most meerschaums made in Europe were three-dimensional, or carved in-the-round, so to speak, the entire bowl and some portion of the shank customarily exhibited bas- and/or high-relief carving. A pipe or cheroot holder fitting this description was, almost always, the creation of European craftsmanship or, alternatively, a European pipe-carver who emigrated and continued to ply his trade in the United States. The three-dimensional pipe is the epitome of artistry and skill. Its manufacture exhibits more care and diligence, more time, energy, and devotion applied to the finished product, signs of a true craftsman's personal pride in his tradework. When compared with European craftsmanship, I am of the opinion that, in general, the American-made antique meerschaum, as in a horse race, can place or show, but hardly ever wins.

Having personally inspected more than a handful of private and public European collections, I found almost none of what I call American-style motifs in either pipe or cheroot holder. What I did find were pipes and holders evoking creative, inventive, fanciful, and original concepts, some outlandish, others bizarre and surreal, and most all premier expressions of fine art. There is also a difference of opinion on record regarding amber mouthpieces that prevailed during the period on both continents. "If a pipe is mounted with a long piece of good amber without a flaw in it, it is pretty sure to be a good pipe. Those of an indifferent quality are seldom carefully and artistically mounted. Of the two kinds of amber, the clear and the clouded, it is advisable to choose the latter. The clouded is tougher, less brittle, and less liable to break in the mouth."[44] "The cloudy amber is the strongest and is the most liked in Europe, the Orient, and in China, while in America, the clear pieces are preferred."[45]

In summary, then, I have presented my thesis regarding approximate dating, and I have offered an unproved concept to determine whether a particular meerschaum, by its motif, might have been made in the United States or abroad. Without the fitted case, I am no better equipped to identify the carver of a pipe or holder than my fellow collectors. A fitted case with a stamped trademark indicates the provenance, but not necessarily the carver, unless one can be sure that the individual named in the case was a sole proprietor engaged in carving meerschaums; the name in the case may not have been the carver's, but that of the retail tobacconist or the exporter/importer.

In the final analysis, if there is no fitted case, who made the meerschaum or in which factory it was produced, or where it was carved, will remain a mystery. If my thesis about dating survives the test of time, at least there now exists some general notion about (approximately) when it was made. But if the meerschaum is cherished because it gives visual pleasure to its owner, this is paramount, and dating and provenance are no longer significant considerations.

Chapter VI. Expositions and Exhibitions Yesterday and Today

As everyone knows, the tenor of the times and the attitude toward tobacco have changed markedly since smoking paraphernalia like those illustrated in this book were made. Beginning almost 150 years ago, commencing with the Great Exhibition at the Crystal Palace in London in 1851, international exhibitions were a platform for pipe manufacturers and those producing related accessories to exhibit their wares to the general public and, particularly, to prospective buyers. "Perhaps the most influential development during this period, to which meerschaum artists were particularly indebted, was the growth of nationally sponsored art exhibitions and international arts-and-commerce expositions." [46] Using a few recorded events, I offer the reader a narrow view into the past when smoking was all the rage, when the exhibition was the forum that gave the pipe-maker visibility and, often, international recognition. This chapter includes not only a brief look into international exhibitions and other exposition forums of the past, but also brings the concept of exhibiting tobacco up-to-date, since these are linked together chronologically with American museum activities of note in this century.

International exhibitions and world's fairs were prominent events that drew global audiences, from as many as 6,039,195 patrons to London in 1851, to as few as 27,722 to Paris in 1889, and 21,477 to Chicago, in 1893.[47] As the first of its kind, the Great Exhibition was very large and very lavish, housed in an iron and glass structure covering 19 acres in Hyde Park. Inside were some 100,000 exhibits—from stuffed animals to recent technical achievements—representing 13,937 exhibitors from all over the globe to demonstrate man's inventiveness and 19th century British achievement in particular. It was, in essence, considered the forerunner of world's fairs. Included in this first catalogued exhibition of the industrial arts of all nations was at least one meerschaum pipe bowl carved by M. Held of Nürnberg, a bas-relief-carved scene of St. George battling the dragon. How were such goods classified? I offer the scheme used at the Centennial Exhibition, also identified as the U.S. International Exhibition of 1876 held in Philadelphia, Pennsylvania. Pipes and associated articles were assigned to manufactures, then assigned to Class 254, further sub-classified as "clothing, jewelry, and ornaments, traveling equipments," and included with "artificial flowers, coiffures, buttons, trimmings, pins, hooks and eyes, fans, umbrellas, sunshades, walking-canes, and small objects of dress or adornment, exclusive of jewelry." [48]

The annual London (Kensington) International Exhibition was an industry trade show noted for always offering a rather English view of the world of objects of art and manufactures, and one in particular, the first held in September 1871, deserves honorable mention. According to the press:

A silver-mounted MEERSCHAUM, by M. HELD, of Nuremberg, representing St. George and the Dragon. We have selected for engraving this out of several drawings sent us by one of the most successful manufacturers of Germany. The article is one upon which much ingenuity is expended; it is often embellished with great skill and taste, and is not unfrequently made costly by the exercise of artistic talent; indeed, a very large proportion of the young Art of Germany is employed in modelling, carving, or decorating these meerschaums. In Germany there are few more productive articles of trade; they are exhibited in the gayest shops; and their ornamentation is generally expensive as well as beautiful.

Rákcózi-style pipe bowl exhibited by the Nürnberg firm, M. Held, at the Crystal Palace Exhibition, 1851. St. George slaying the dragon was idealized in the paintings of Carpaccio, Orsi, Rafael, Rubens, Tintoretto, and Uccello, and in a Sir Joseph Edgar Boehm sculpture. Here, the patron saint known to Beirut, Lebanon, and Great Britain, a Christian martyred under the rule of the Roman Diocletian in 303 A. D., is captured in meerschaum, 6" h. *The Great Exhibition. London 1851.* Facsimile, Crown Publishers, Inc., 1970, 17.

Here Messrs. Charles Goas and Co., of 23, Finsbury Place, London, make a good exhibition of Viennese pipes, manufactured by M.A. Nowatry, of Vienna. Some of the meerschaums are large, and represent portraits of distinguished persons, executed with that finish characteristic of paintings of Continental masters. Many of them manifest an extraordinary ambition of design—one representing a Cabinet Council, with her majesty sitting on her throne, and Ministers in consultation at the table.[49] There are some curled meerschaum cigar-holders of great richness of design. Mr. Franz Hiess, of Austria, models his meerschaums after animals; they are very ingenious and executed with great force. Messrs. Ludwig, Hartmann, and Eidan [sic] also show a variety of massive meerschaum pipes, in which the mere bulk of meerschaum and amber presents richness and distinctness of appearance. A. Trebitsch, of Hungary, has also an attractive collection adjoining those we have named, the pipes including designs of both animals and buildings. One meerschaum presents quite a little farmyard of animals, with chanticleer in a very proud position upon the bowl.[50]

In 1873, a writer for *Practical Magazine* reviewed the third annual London International Exhibition:

The meerschaum pipes are varied, curious, and some of them very costly. One, the bowl surrounded with an elaborately carved group of figures, is priced at no less than a hundred and fifty-five guineas. Of course, no one would think of smoking such a pipe, except for the name of the thing; it would be more trouble than pleasure for general use. Another displays an ambitious attempt to represent the light cavalry charge at Balaklava. The troopers are not very good representatives of British hussars and dragoons, and it is not quite clear which side is getting the worst of it; but much ingenuity is certainly displayed in carving the meerschaum into such forms. The way in which some of the bowls are tortured into representations of heads of animals, or of human beings, is more curious than beautiful; while in other cases a few touches by the knife or chisel of the carver have sufficed to produce a really graceful and artistic design.[51]

Another reported:

Meerschaums are frequently mounted in silver, and have sometimes been decorated with jewels, so that their cost has been excessive. They are generally enriched with ornaments in high relief, executed with much beauty, and embracing a vast variety of design. The care with which this material may be moulded and fashioned by the artist (for such he

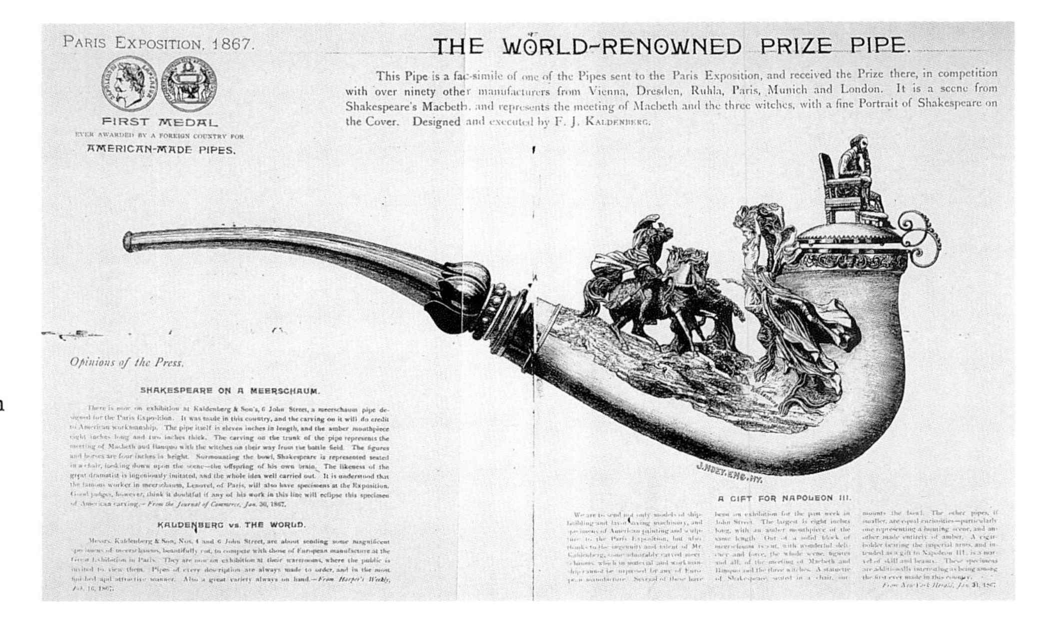

PARIS EXPOSITION, 1867.

FIRST MEDAL

EVER AWARDED BY A FOREIGN COUNTRY FOR

AMERICAN-MADE PIPES.

THE WORLD-RENOWNED PRIZE PIPE.

This Pipe is a fac-simile of one of the Pipes sent to the Paris Exposition, and received the Prize there, in competition with over ninety other manufacturers from Vienna, Dresden, Ruhla, Paris, Munich and London. It is a scene from Shakespeare's Macbeth, and represents the meeting of Macbeth and the three witches, with a fine Portrait of Shakespeare on the Cover. Designed and executed by F. J. KALDENBERG.

Opinions of the Press.

SHAKESPEARE ON A MEERSCHAUM.

KALDENBERG vs. THE WORLD.

A GIFT FOR NAPOLEON III.

Facsimile of the Kaldenberg pipe that received the Paris Exposition medal in 1867. The scene is of Macbeth, Banquo, and three witches, with a small statuette of the contemplative Shakespeare as a finial atop the meerschaum windcap, 16" l., American, ca. 1867. F. J. Kaldenberg Co. Catalog, ca. 1893, 32-33.

This pipe, made by August G. Fischer of Orchard Park, New York, for the Pan-American Exposition of 1901, depicts our Native American stalking a buffalo, and separating the two is a sphere symbolizing the globe, 8" l., 4" h., American, ca. 1900. *Courtesy of the Author's Collection.*

> is), who decorates the bowl, allows the greatest ingenuity and elaboration of design to be exhibited in this branch of art-manufacture. Most pipe-sellers and tobacconists can exhibit specimens which are perfect miracles of patience and labour, and are worth forty or fifty pounds each. They are enshrined in velvet, and shown like jewels of a Marchioness.[52]

Yet another witnessed all the sumptuous pipes in the exhibition. Next to a stone pipe from Finland was

> a noble Russian bowl of yellow meerschaum, standing five inches high, and carved with the lineaments of a grandly rugged countenance, such as bespeaks the Northland patriarch; and the stem of this princely pipe is formed of twenty various kinds of precious stones, finished by the cunning hand of the lapidary.[53]

One exhibitor of merit was F. J. Kaldenberg of New York City who claimed to have introduced the meerschaum and amber industry to America. The company's first exhibition was at the American Institute in 1865. This company had become world-renowned when it sent a series of pipes specially designed to the Exposition Universelle in 1867. The company was to compete with more than 90 other manufacturers from Dresden, London, Munich, Paris, Ruhla, and Vienna. Kaldenberg sent 30 of his specimens to this exposition and became the first American pipe company to be honored by a foreign country, the recipient of a bronze medal (gold being reserved for inventions) for "best meerschaum goods." Before officially receiving the medal, however, he was required to furnish a number of affidavits to prove that the pipes were made entirely in the United States.[54] Among the 30 submitted for the competition was a pipe expressing Macbeth's encounter with the witches on the heath.[55] In Paris, a gentleman offered to purchase this pipe for 10,000 francs and the offer was rejected. As recorded in a Kaldenberg Company catalog and quoted in the press:

> We are to send not only models of ship-building and labor-saving machinery, and specimens of American painting and sculpture to the Paris Exposition, but also thanks to the ingenuity and talent of Mr. Kaldenberg, some admirably carved meerschaums, which in material and workmanship cannot be surpassed by any of European manufacture. Several of these have been on exhibition for the past week in John Street. The largest is eight

inches long, with an amber mouthpiece of the same length. Out of a solid block of meerschaum is cut, with wonderful delicacy and force, the whole scene, figures and all, of the meeting of Macbeth and Banquo and the three witches. A statuette of Shakespeare, seated in a chair, surmounts the bowl. The other pipes, if smaller, are equal curiosities—particularly one representing a hunting scene, and another made entirely of amber. A cigar-holder bearing the imperial arms, and intended as a gift to Napoleon III, is a marvel of skill and beauty. These specimens are additionally interesting as being among the first ever made in this country.[56]

At the 1876 Centennial Exhibition in Philadelphia, clay, meerschaum, and amber goods manufacturers and pipe trimmings from the Austria, Canada, Germany, Great Britain, the Netherlands, and United States and were represented in large numbers.[57] For this exhibition, Kaldenberg arrayed some 1,500 different stock pipes, and offered the public the very first view of his master creation, "Columbia," a 28-inch high hookah-style meerschaum pipe resting on a plinth. This massive sculpture depicted four high-relief-carved figures at the base representing agriculture, commerce, manufacture, and navigation, four intermediate-level cherubs representing music, painting, literature, and sculpture, and at the very top, Columbia, representing power, justice, and liberty.[58] Anecdotally, Kaldenberg was commissioned to carve the likeness of the wife of Major General Ruffalorich, a retired Russian Army officer living in New York City at the time; her portrait was used as the image, and the finished pipe, 15 inches in length, with an amber mouthpiece was, in its day, a cost of $150![59] Rather pricey for its day, don't you agree? Not at all, according to this *New York Herald* reporter who also commented that "there is one now in this city, which, for its graceful form and elaborate carving, is valued at five hundred dollars."[60]

The William Demuth company had exhibited its best pipes at the American Institute Fair in 1869 in New York, at the Philadelphia Centennial Exposition in 1876, and at the World's Industrial and Cotton Centennial Exhibition in 1884 to 1885 in New Orleans. I am convinced that just as Macy's and Gimbels had been New York City's fiercest retail competitors for many years in this century, Kaldenberg and Demuth were the city's keenest rivals for the consumer's tobacco dollar in the last half of the 19th century. At the World's Columbian Exposition in 1893 at Chicago, in which smoker's articles were prominently displayed, Demuth presented the company's best meerschaums that had been assembled to that date, and he exhibited the colossus "Columbus Landing in America" pipe to vie with Kaldenberg's "Columbia" at Philadelphia 17 years earlier.[61]

During this 50-year period, one can find an occasional meerschaum that received a competitive award at one of these world fairs. Altogether, some 180 regional and national fairs were held from 1851 to1901, but only the most significant exhibitions are included in the table at Annex B.[62]

Another August G. Fischer masterpiece depicting Native American lore made for the Pan-American Exposition, three mounted warriors chasing three buffaloes, 19.25" l., 3" h., American, ca. 1900. *Courtesy of the SP Collection.*

Other venues of a more localized nature must also be mentioned because they, too, have had great consequence for the pipe collector. The two most noted exhibitions of the 19th century were arranged by William Bragge, F.S.A., a master cutler of Birmingham and Sheffield, England. His recorded collection numbered about 7,000 specimens of art and industry from every quarter of the world and nearly every time period. Because his greatest interest was ethnographic pipes, the strength of his collection was in tribal-art pipes. However, in his early years, he did possess 25 "old carved meerschaum pipes." A two-day exhibition at the Birmingham and Midland Institute, Birmingham, England, on December 15 to 16, 1870, demonstrated to all who visited that "the history of pipes is the history of man."[63] Ten years later, the Bragge Collection was on loan to the Edinburgh Museum of Science and Art, and a visitor's guide to the Collection was published. "In [case] No. 22 are some massive and finely carved meerschaums, among which mention may be made of a mouthpiece with Tyrolese peasants and dog in full relief."[64]

The Demuth Collection, by now considered one of the most complete and comprehensive found anywhere in the world, was present at the Exposition Universelle in 1900 at Paris, and on view at New York's American Museum of Natural History for a brief interval in 1905. On March 26, 1952, the Rochester Museum of Arts and Sciences sponsored a one-day exposition of artifacts from the American Tobacco Company, "The Story of Tobacco: An Exhibit Showing Its History, Growth and Use." In the vitrines were American Indian smoking artifacts and the prized American Tobacco Company's "Half and Half Collection" of antique pipes, including Demuth's Columbus pipe and the matched miniature meerschaum busts of 29 presidents of the United States.[65] The Columbus pipe is now ensconced in the Austrian Tobacco Museum, Vienna, and the singular set of presidential busts is in a private American collection.

Antiquarian tobacco artifacts have gained a modicum of respectability in the last 15 years, at least in the opinion of those responsible for organizing these events. Yale University's Peabody Museum of Natural History, New Haven, Connecticut, featured a quantity of stone effigy pipes in "Smoking to the Gods," in 1985. In 1986, the R.J. Reynolds Company commemorated the dedication of its Tobaccoville Manufacturing Center in Winston-Salem, North Carolina, with an exhibition of early advertising art titled "Golden Leaves." In June 1987, the Museum of American Folk Art in New York City sponsored "Tobacco Roads: The Popular Art of an American Pastime," an assembly of some 70 personal-use tobacco objects. From April to October 1990, the Valentine Museum, Richmond, Virginia, offered "Smoke Signals: Cigarettes, Advertising and the American Way of Life," which included mixed media artifacts depicting Richmond's tobacco industry through advertising. In June of the same year, the Yale Center for British Art, New Haven, Connecticut, offered "Powder Celestial: Snuff Boxes, 1700-1880," a tribute to snuff mulls, boxes, and flasks. The Historical Museum of Southern Florida, Miami, sponsored "Tobacco Art: Cigar and Cigarette Labels From Cuba and Florida," from October 1995 through January 1996. "The Image Business: Shop and Cigar Store Figures in America" was a comprehensive traveling exhibition of more than 60 figures on view at three different east coast museums between May 1997 and April 1998. Perhaps the most celebrated research site for tobacconists is the New York Public Library, which has housed the George Arents Jr. Tobacco Collection for almost a half a century. From September 1997 to January 1998, "Dry Drunk: The Culture of Tobacco in 17th- and 18th-century Europe," was the talk of the city, the state, and the country. A selection of rare drawings and images, books, and related artifacts from the Collection was on public display for the first time since 1960 when Arents willed it to the Library.

Only one of the aforementioned exhibitions focused on pipes, and that is why, until mid-1998, no other museum in this country was more important to antique pipe collectors than the Museum of Tobacco Art and History, Nashville, Tennessee. David R. Wright was the curator of this privately funded, open-to-the-public museum on the premises of the United States Tobacco Company. The museum opened in July 1982 and rapidly established itself as a premier locus for everyone interested in any aspect of tobacciana collecting: pipes; tobacco jars, boxes, and tins; snuff jars, bottles, boxes, and mulls; pipe tampers; cigar accessories; cigar-box labels, caddy labels, advertising posters, billheads and mastheads, and other advertising ephemera; and, saving the best for last, a superior collection of cigar-store Indians and other tobacco figures. In a word, the Museum had a permanent display of all the beautiful objects related to tobacco use that were made during the last 400 years. David always promoted a continuing exhibition program, and he offered the public at least one major exhibition each year on tobacco art, collectibles, or antiques. Although out of chronological sequence with the exhibitions mentioned previously, the most memorable event in recent history for the American meerschaum collector was "Meerschaum Masterpieces: The Premiere Art of Pipes," an exhibition that ran from November 15, 1990, to February 23, 1991. More than 100 exquisitely crafted pipes and holders from private collectors were on view, and to ensure its posterity, in conjunction with the debut, an illustrated catalog bearing the exhibition's title was published. Sadly, in July 1998, in a less-than-sweet-sixteen anniversary announcement, the Museum precipitously and unexpectedly closed. Today, American collectors have no public museum, and those who had ever visited Nashville know how much we will miss David and this singular center of antiquarian tobacciana that he had mobilized.

However, a few outposts for antiquarian tobacco treasures survive. Antique pipes can be seen in a few traditional retail tobacco shops where yesteryear tobacco utensils are displayed alongside today's retail products, and just about every pipe collector knows these shops well. A list of retail tobacco shops and museums in both the United States and Europe is found at Annex C.

Part II: The Collector's Market Today

Chapter VII. Valuation: How Much Is It Worth?

At the request of the publisher, a price guide is included in this book. One may question the utility and appropriateness of a price guide in an art book, so I desire to put the guide in its proper perspective. A price guide offers broad, general, and conservative guidance on the current value of an item to seller, buyer, and personal effects appraiser. The values in a price guide may be presented either as a single dollar amount or as a value range. When a price guide is used in isolation, the values are really sterile numbers without relevance, so the reader should view even the contents of this price guide with caution. A price guide is also temporal, since every item has its day, and investment potential fluctuates in relation to a generation's changing taste for antiques. Let me explain.

The first general law regarding antiques is that value is not governed by date. This is especially true for meerschaums. In some fields of collecting, older equates to more expensive, or greater value, but in this collecting field, older is not necessarily more costly. In Europe, today, the tendency is to prize meerschaums from the first half of the 19th century because they have historical significance. Ask any American meerschaum collector, and he'll admit that he rarely spends much for an early 19th century lap pipe with a cherrywood push stem, because they're not in great demand in this country; he'll pay considerably more for a late 19th century ornately carved European-produced meerschaum pipe with an amber mouthpiece and a fitted case.

The current-market price of any antique is derived from the evaluation of many critical factors, such as size, condition, relative rarity of the motif, and the degree and amount of ornamentation; collectively, these factors determine the value. A dictionary definition states that value is the worth of something in terms of the amount of other things for which it can be exchanged, and valuation is estimating the item's value or worth. The two terms are intertwined and interdependent. For the appraiser, valuation is the act of determining estimated or fair market value, a term most often used in formal appraisals. Valuation goes beyond setting a monetary value, because it describes an object with as many physical details as possible, and it represents a professional opinion about the item's quality, authenticity, and design. Valuation, a judgmental, imprecise, and interpretive science, attempts to weigh many diverse tangibles carefully, because the appraiser must consider price trends over time, traffic at the auction block, and other relevant factors for objects of a similar caliber, a procedure also known as market analysis. The valuation estimate can be different for each type of appraisal, for example, one performed for replacement, or fair market, marketable cash, or liquidation, because each serves a different purpose and, more often than not, the valuation yields a value range rather than one of exact dollars and cents. In a word, valuation is a subjective procedure of assigning a price to an item which requires looking at it, knowing what it is, and what it is worth.

The estimated value is the summed dollar total for all the factors that account for an object's worth: quality, original condition, age, and authenticity, but current- or fair-market value is determined only when both a motivated and informed seller and buyer are prepared to transact at a given price; when object and money have changed hands, when title has passed from seller to buyer, the current-market, or fair-market, value has been derived, at least for that item and only at that precise moment. Thus, the fair-market value is not an educated guess, not a considered opinion; it is, in the opinion of the final authority—in the United States—of the Internal Revenue Service, the price that property would sell for, or the actual price it has sold for, on the open market.

Jane Brennan, once the curator of the U.S. Tobacco Museum, Greenwich, Connecticut, offered a simple checklist to guide valuation of an antique meerschaum:

> Is the pipe complete?
> Is the stem made of amber?
> How old is the pipe?
> What is the pipe's condition?
> What about the carving?
> What about the color?
> How big is the pipe?[66]

I believe that the valuation of a meerschaum is a trifle more complex that the collective answers to these seven questions, and I offer my rationale for this view. While most understand that quality, condition, and age are components in the valuation of any antique, for a meerschaum, additional criteria, such as execution and rarity, must also be considered. I dispense with execution first. The detail of an antique meerschaum pipe is extremely important, but comparable detail (or equivalent ornamentation) on a cheroot holder should also warrant a high value, sometimes a greater value, because a smaller piece is much harder to manipulate and carve. Intricate detail on a small object signifies the carver's exceptional skill and diligence. Moreover, because a cheroot holder or a cigarette holder is more fragile than a standard-size pipe, all other things being equal, if both a pipe and a holder of equivalent stature display no visual chips, flakes, fissures, or breaks, the holder may command a comparable value, perhaps as high as that of the pipe. Of course, I realize that this type of comparison cannot often be made; usually, the piece in question must stand on its own artistic merits.

The key to fully comprehending valuating this type of art is that execution and, to lesser degree, rarity of motiv—not size—should dictate the price. Rarity is an elusive attribute because only the experienced collector knows whether a particular motif is common or scarce. I previously mentioned the hand with egg and the talon with egg; these two motifs, horses, dogs, deer, and Nubian, Bedouin, and school ma'am (sometimes confused with the more elaborately carved Gibson girl) figuratives, however well executed, are not rare motifs. Rarity should also not be based on the overworked characterization of a particular motif as being "one-of-a-kind." No meerschaum collector, here or abroad, is sufficiently familiar with the entire output of this cottage industry to make this claim. Then what are rare motifs? I would say that mythological characterizations, such as Venus, Diana the Huntress, Pan and naiads, Mephistopheles, celebrated battle scenes, intricate architectural statements, anything naïve, suggestive, and erotic (the pornographic motif is mentioned in the next chapter), the odd, the peculiar, even the bizarre, are all premier motifs. The visual effect of just about every talon, hand, deer, and dog pales in comparison to a fascinating and unusual expression of subject matter executed with a sense of whimsy or fantasy that stimulates all the senses, challenges the mind and, above all, represents ingenious creativity.

It is not uncommon to find similarities among, for example, Nubian, pasha, and Gibson girl figuratives, as well as many other popular motifs carved at various centers in Europe for export to, and those made in, the United States. The often-heard comment, "I have one exactly like that," is not quite an accurate statement, since every handmade object, by definition, has ever-so-slight differences in production, one to another. However, the statement "I have one that is similar in style and size" might be more accurate. Meerschaums of this era were not mass-produced, but they were often copied—some very well and some poorly. I have purposely included illustrations of a few "variations on the theme" to offer evidence that multiple expressions of some popular themes were produced, but they were not necessarily from the hand of the same carver, or from the same manufactory, or even from the same carving center.

What about size? While bowl height will vary markedly from meerschaum to meerschaum, pipes measured, on average, from 5"- 9" in length, and cheroot and cigarette holders measured, on average, from 4"-7" in length. Many pictured in this book exceed these average dimensions, but, I believe that these were not made to be smoked. I contend that these larger, outsized meerschaum pipes and holders were carved to celebrate, to commemorate, or to glorify an event; to compete at world's fairs and exhibitions; or to commercialize, that is, to be used as window dressing or as a retail tobacconist's shop sign. I have no proof to support this claim, but after studying the social custom of smoking in the last century, I cannot find a practical use for a pipe or holder that measures more than 10". Such a meerschaum is, more than anything, a conversation piece that prompts onlooker "oohs" and "ahs." The market price for a meerschaum of this caliber and stature is, not surprisingly, markedly higher than that for an average-size pipe or holder; however, finding one is not an everyday occurrence.

In assigning value, condition and completeness cannot be completely ignored. The original condition and completeness of the meerschaum bowl and shank are critical factors. Damage, however inconsequential or insignificant, should not be minimized. By definition, completeness includes the mouthpiece, and the original amber mouthpiece is not a trifle. Raw amber, today, is extremely expensive, and amber formed and fitted exactly to the diameter of the shank of the pipe or holder is yet more costly. Its absence devalues the pipe, but the degree of devaluation is very subjective. In my opinion, the primary emphasis of "condition" is on the meerschaum, not the mouthpiece. Any skilled pipe repairman can replace an amber mouthpiece with one of composition, bakelite, or other material, but a diligent art restorer able to reconstitute damage to a meerschaum is less easily found. The fitted case adds a small amount of value because it might identify the carver, and it serves to protect and preserve the meerschaum.

Perhaps the best words to describe the way in which to understand the science of valuating an antique meerschaum is to quote a craftsman who began carving in the 1880s and whose interview was published on July 30, 1938, in the Vienna, Austria, newspaper *Völkischer Beobachter*. He told the reporter: "There were pipe-carvers and pipe-makers and that the difference in quality between their work approximated pretty well to that between Michaelangelo's [sic] and a house painter's."[67] Quality dictates the price in any commodity, and in these uniquely crafted expressions, quality is all-important, a characteristic that must be applied not only to condition, but also to execution, size, and rarity of motif, in order to derive a current-market value. In a word, not all antique meerschaum pipes are valuable, but some antique meerschaums are priceless. The trained eye and a lengthy experience in collecting are the keys to discerning the difference between the meerschaum work of a Michelangelo and a mechanic. When the collector can discern the difference between the works of both, he has arrived.

In closing, this price guide is my rough-order-of-magnitude estimate of the current-market value of each piece illustrated in this book. These estimates are based on 40-odd years of hands-on experience—what I have seen in museums and private collections around the globe, continuous research, and recent prices paid at auction and antique shows. This guide is not, however, designed to achieve consensus; it is my valuation, and it should be used as broad, general guidance. The owners of the meerschaums illustrated in this book, other pipe collectors, and antiques dealers may violently disagree with the assigned market values, but remember that a price guide constitutes only one of many factors governing the market value of an antique meerschaum. I leave to both seller and buyer to mutually determine what to pay for an antique meerschaum. No doubt, both will attempt to use this price guide to govern the transaction, but I hope that I have been convincing in my argument that, used in isolation, the price guide is an inadequate yardstick of value. In the end, the price paid for an antique meerschaum will usually be derived using some magic mental algorithm or imagined formula that results in a price to which both seller and buyer can agree, the amount that the buyer is willing to expend for what he believes is a meerschaum "to die for."

Chapter VIII.
Searching for Holy Grails

Is it possible to start collecting antique meerschaums today? Are pieces like those exhibited in this book still available, and are they readily found? The answer to these questions is "yes"! Beyond having an elementary understanding of this collecting field, what else does it take to be one of us? The prerequisites are focused persistence, vigilance, diligence, commitment, networking, timing, lots of luck, and some disposable income, but not much else. Follow the standard sources that just about every other antique collector uses: visit the local antique fair and flea market; subscribe to magazines and trade papers specializing in antiques; and maintain contact with those auction houses known for occasional sales of antique pipes. These are basics essential to staying alert and active in any hobby. The newest source, and keen competition to the auction house, is the Internet's worldwide web on-line auctions, or E-commerce. Those who are accustomed to Internet "surfing" or "trolling" have already found several web sites catering to both pipe smokers and pipe collectors. One, the San Jose, California-based e-bay web site, offers a wide range of antiques and collectibles with on-line bidding by e-mail. Cyber contenders include eHammer.com, Antiques.com., ForSale.com., AuctionsUnlimitedInc.com., Skybid.com, and Auctionuniverse.com. Given the exponential expansion of the Internet and the increasing popularity and success of E-commerce, I suspect that by the time this book is published, many more Internet auction sites will be on line.

Here are a few fundamentals to consider:

- I estimate that the median age of the American antique pipe collector is at least 50. Although the thought is rather morbid, it is fact that their collections will probably be dispersed some time in the future through private sale or the auction house, and their meerschaums will be in circulation once again.
- Collecting antique meerschaums seems to attract median (or higher) income classes. This field is not for the economically faint of heart. Great "stuff" always commands big prices![68]
- Demand remains slightly greater than supply. I say slightly because not every collector of antique pipes is interested in antique meerschaums.
- When I began collecting in 1959, information about antique pipes was almost non-existent. Today, specialized books, antique pipe collectors, and museums possessing antiquarian tobacco artifacts are much more accessible.
- A visit to one or more of the specialized museums will serve you well. If you are inclined, go to Paris, or Vienna, or just about any city in Germany, and visit many of the continental European tobacco museums. Their holdings are spectacular, and the experience is mind-expanding.

• As the new millennium approaches, I believe that American anti- and non-smoking forces will pursue a more aggressive campaign to eliminate tobacco as a consumer commodity. As with all other artifacts destined for extinction, they then become popular to collect. Condom tins, Moxie, cap guns, walking sticks, vinyl LPs, and a host of other items of our past that have fallen into disuse are now or are becoming "hot." Heightened collecting activity generally follows and usually draws things out of closets and attics. I am convinced that the closets and attics of many American homes are caches for antique meerschaums.

• We collectors have never seriously considered organizing a national society, publishing a journal, or establishing a clearinghouse for information exchange. Because each collector is left to his own devices, the neophyte should befriend a meerschaum maven and learn from him.

• In the last 20 years, auction houses here and abroad, particularly Christie's South Kensington, London, have included meerschaums in sales of "varia," "objects of vertu," and "Victorian and Continental Miniatures." Rare meerschaums do continue to find their way to the open market with regularity.

• The pipe show is an excellent venue for access to collectors and opportunities to buy, sell, trade...and learn.

The pipe show, a concept that is almost 20 years old, is a unique source for the meerschaum collector. The first national pipe collectors' exposition, sponsored by Drucquer & Sons, Ltd., of Berkeley, California, took place in Burlingame, California, in October 1980. Since that year, regional and local pipe clubs across the United States have sponsored similar shows with varying frequency. Not all pipe shows include exhibits of antique pipes, but one never knows. That's where networking plays its part. Getting to know collectors and learning when and where antique pipes will be exhibited is the most practical way to stay in touch with all the active players.

The key to building and enjoying a collection is education. Ask any serious collector and he'll tell you that knowledge is power, the power to discern quality and rarity in that one meerschaum among many; to know when, where, and from whom to buy; and to recognize a fake, a Turkish-made facsimile, a restored piece, and a hydrostone reproduction. The knowledgeable collector also commands that ultimate power, the power to resist buying impulsively. Knowledge can only be derived through continuous reading and research. Buying a book before buying the first meerschaum is fundamental. Even our "closet" collectors who prefer to remain anonymous rather than maintain contact with mainstream collectors, would willingly admit that their most important tool is a good reference library. That library should be comprised of not only the basic specialized books, but also those that keep abreast with new scholarship. Auction catalogs are also an important reference because they help to trace the provenance of select pieces and provide a sense of current market values. On April 4, 1978, Sotheby Parke Bernet, New York City, auctioned an outstanding collection of carved meerschaum and briar pipes, Oriental pipes, water pipes, and related material. The auction catalog, "Sale 615, Pipes," and the accompanying finalized price list, were used as a benchmark for values at this time as an excellent gauge of market price trends.

After networking, some serious reading, and visits to a few pipe shows and a museum or two, it is time to purchase an antique meerschaum. Here are some of my thoughts, listed in no particular order, for consideration:

• Buy the best pieces you can afford that are as close to their original state as possible. The cost for replacing an amber mouthpiece is prohibitive, and almost no one, today, repairs meerschaum.

• If you believe that a particular meerschaum is the best of class, a premier piece, buy it, because these opportunities are infrequent.

• Be discriminate, buy only what you like...and decide quickly.

• On average, prices for antique meerschaums have been relatively steady-state for the last 20 years.

• If you are about to make a purchase, but are not fully engaged or 100 percent convinced, repeat my favorite phrase: "there's always tomorrow."

• It is more prudent to purchase five $1,000 meerschaums than one $5,000 meerschaum. Only a few collectors aim this high. If you later decide to dispose of your collection, a larger audience awaits the lower-priced pieces.

• Never miss the opportunity to appreciate the aesthetic qualities that make every antique meerschaum important, especially those that belong to others. Learn from looking, and gain knowledge from close inspection of what makes each extraordinary or notable.

• Look for artistic nuances that separate the extraordinary from the ordinary, such as:

—Inset glass eyes in figurative heads of humans and animals.
—Scarlet red amber mouthpiece, the most luxurious of amber colors.
—Highlighting, or the absence of beeswax, a carving technique known as bi-coloration that effects milky-white accents on the meerschaum's surface.
—Chased, engraved, etched, or repoussé silver or gold shank band.
—Inset or encrusted semi-precious or precious stones.
—Reticulated silver or silver-plated windcap.
—Traceable silver hallmark.

• With very few exceptions, antique meerschaums of this period never bore the carver's signature. If you find an incised signature on the shank or underside of a meerschaum, it's probably a fake or Turkish.

• Light, air quality, temperature, mold, and pests have deleterious effects on meerschaum, so use care and preserve them, however small the collection.

As to fakes and Turkish meerschaums, permit me a few words so that I may segue into something less pleasant to mention. Some 200 years ago, pressed, chip, or mere-sham pipes and holders were produced, advertised and sold as such, and distinguishable from those made of block meerschaum. Nearing 1900 and until about 1930, mail-order houses, such as Montgomery-Ward, Sears Roebuck, and the Hudson's Bay Company, offered not only genuine block meerschaums, but also "imitation" and "second-quality" meerschaums, carved and plain, at a cost of between $1.95 and $4.95. The majority of these "imitations" were pressed meerschaum lap-style pipes with cherrywood stems and horn mouthpieces, identified by exotic names as "German antique pipe," or "The Fatherland Antique Vienna Meerschaum," and the bowls typically had an incised false date, such as 1710 or 1725. I have to assume that anyone who bought such a pipe knew exactly what was being advertised. Therefore, I must take issue with the following that labeled this type of pipe as fake:

> One elaborate fake in meerschaum pipes is made up of a heavy bowl, sparingly and somewhat crudely carved, and fitted with a mouthpiece decorated with mother-of-pearl. Pipes of this kind were produced in quantity in about 1890, and were made from meerschaum parings molded and pressed together. Often a spurious date, up to one hundred years earlier, is incorporated in the carving.[69]

For about the last decade, I have been aware of facsimiles of antique meerschaums being produced in and exported from Turkey, but these are block meerschaums that have always been consistently advertised and marketed in this country as Turkish. To the best of my knowledge, Turkey does not produce pressed meerschaums. So what about the art of deception, the concern for the proliferation of fakes? Every collecting field has its share of charlatans, and someone or something always comes along to impact a collecting field negatively and spoil the fun. Recent events suggest that someone and something are now taking their toll on this field. Pseudo-antique meerschaum pipes are appearing on the auction block and at the antique fair in Europe and in the United States. These fakes—or forgeries, or frauds—are cast of an epoxy resin, are being produced in quantity, and are circulating far and wide. At least two discrete pipe motifs have been identified to date, and there is ample evidence that both were produced in mass by the same unknown source. One motif is a group of high-relief-carved carousing merrymakers circumscribing the bowl, one man in a top hat adjacent to several women; the entire pipe is uniformly pre-colored in an oxblood finish, and a reticulated, tapered, pot-metal shank ferrule connects the bowl to an orange-amber-colored bakelite mouthpiece. Two clues that this pipe is a fake are, first, a visible vertical seam on the bowl, indicating that the pipe was cast in a mold and, second, the incised scripted signature on the mouthpiece of Gustave A. Fischer, someone who never signed his pipes! On the shank of this pipe appears the incised signature, "Gustav Fisher Jnr," and "Gustav Fisher, Vienna, Exposition 1873" is engraved on the mouthpiece. The other motif is also a uniformly pre-colored oxblood finish tulip-shaped pipe bowl with a high-relief-carved female nude on the front. It, like its fake companion, has the same ferrule and the same type of mouthpiece.

One of two fake pre-colored epoxy pipes now in circulation. *Courtesy of the SP Collection.*

These two pipes are chimerical fakes because: (1) the last name of both families is spelled Fischer, not Fisher; (2) Gustave A. (of Orchard Park) was an employee of the William Demuth Company until 1892, so it is unlikely that he would have been permitted to sign a pipe while working for this company; (3) Gustav Fischer Sr. was still in Vienna in 1873, and he could never have carved such shoddy pipes. If these had been made by Fischer for this exposition, I would submit that the incised stamp would be the German *Weltausstellung Wien*, not Vienna Exposition; and (4) his son, Gustav Jr., was born 14 years after the Vienna Exposition. The very first example of this latter variety was scheduled for auction at Christie's South Kensington, London, in late 1994, but it was withdrawn after a representative from Christie's Department of European Works of Art questioned its authenticity. Sadly, the prices of both at auction have approximated those normally attributed to fine-quality antique meerschaums. *Caveat emptor!* Be on the lookout for these two, and, for all I know, others of this genre are now in circulation.

A few more words to the wise. Most meerschaum collectors I know are in it for the long haul. They buy because they want to possess, to retain, not to resell. I have seen a fair number of "buy low, sell high" and "get-in, get-out-quick" collectors come and go these many years. None of these has ever made his first million by selling to knowledgeable collectors. Their tactics are patently obvious, their knowledge base is shallow and, generally, the quality of their offering is poor. Be wary of this type of shake-'n'-bake expert, although he may speak with authority and appear to have a command of this craft. Stereotypically, he is famous for pithy phrases such as "you'll never see another meerschaum like this one again"; "it's a limited edition, signed piece"; "I've never seen anything like this before...ever"; and "although the case is missing, take my word for it, Czapek carved this one." I have been studying this collecting field for almost 40 years, my library consists of almost 4,000 books, catalogs, and pamphlets on tobacciana—all of which I have read—and I am still unable to make such categorical statements.

As a final word, anyone interested in collecting antique meerschaums has to start somewhere. This book will go a long way to making every beginner antique-meerschaum literate and current-market-value smart.

Part III: Meerschaum Retrospective

The balance of this book is devoted to illustrating a vast array of rare and unusual meerschaum pipes, cheroot holders, and cigarette holders. The illustrations are grouped thematically into chapters, and similar objects are placed side-by-side without regard for their size. A substantial segment of a typical antique meerschaum collection is figuratives, because they are popular and easily acquired. Figuratives can be impressive, historic, and colorful additions to a collection. However, because illustrations of figuratives typically dominate just about every book and exhibition catalog on antique pipes in circulation, I selected seven domains in which figuratives play a minor role. These seven domains are celebration and commemoration; femmes fatales; of mice and men; myth and fable; naiads, putti, cherubs, and seraphim; whimsy, fancy, quaint, and curious; and naïve, suggestive, and erotic. Out of range of this book are pornographic meerschaums that were also made in significant quantities in the last half of the 19th century, but the vast majority were crudely carved, and the subject matter was almost always distorted or exaggerated. Middle-class respectability and acceptable social custom dictated that the well-bred Victorian man use such a pipe or holder at the end of the meal and only after the women retired.[70] Pornographic motifs might titillate the viewer's senses, but none of these lusty expressions is included, because I am of the opinion that naïve, suggestive, and erotic meerschaums leave more to the imagination.[71]

Why these particular domains? After careful and protracted study, I chose them because I believe that each will contain some of the most striking examples of the craftsman's art. These seven domains do not, by any stretch of the imagination, account for the totality of all the topical themes that were probably produced during that time period, but, in my considered opinion, the very best in meerschaum craftsmanship has been expressed in one or more of these domains. Carvers evoked myriad other motifs and messages, but the meerschaums that draw the attention, admiration, and awe of the viewer seem to fall into one of these seven categories. Without question, figurative subjects—man, woman, and all of God's creatures—were carved in quantity, but however ornate or intricate they may be, the coordination of hand, tool, and eye, in my judgment, could never be as demanding as that required to craft a high-relief group of frolicking children at play atop a cheroot holder, or a pipe with a to-scale image of Diana, the virgin goddess of the moon and of hunting, astride her steed. Balance, proportion, and the powerful *contraposto* of a three-dimensional scene was, by comparison, exceedingly more difficult to execute than, for example, an anatomically correct life-like bust of a man with mutton chops wearing a kepi, or a goateed soldier wearing a shako.

Since I chose these domains, I take responsibility not only for excluding a representative number of figurative subjects—against the wishes of a few contributors—but also for the many hundred extraordinary pipes and holders that merited serious consideration and have been judged as distinctive, but were eliminated during the vetting process. I also assume full accountability for the organization of the illustrations. No doubt, some readers will question why I placed a particular piece in one chapter, rather than in another. What is important is that the piece is included for your viewing pleasure. Finally, I acknowledge that each of the following chapters is not given equal illustrative prominence. My objective was to select the best of the best, motif notwithstanding.

In the succeeding pages, however, the illustrations of a few select figuratives are included, not because they are the best of class, but because they are rather unusual. As well, several expressions that are almost alike but, simultaneously, different in their appearance and articulation are illustrated. The reason for incorporating these is obvious, to demonstrate how various carvers elected to use their individual imagination to express what is apparently the same motif. Finally, I applied great diligence to select a broad and representative assortment of subject matter in both pipes and holders for each domain, and all of these, to the best of my knowledge, have their first public appearance in this book.

The names of some of the donors are shown in the following list and, for economy and ease of identification, only their initials are used in the captions. Others who participated have elected to remain anonymous, and I have honored this request. Their contributions are acknowledged as the property of private collections. To contact any of the listed donors, please write to me in care of the publisher.

Name	Initials	Name	Initials
J. Trevor Barton Collection, England	JTB	Mark Goldman, Mom's Cigar Collection	MG
Dr. Federico Baylaender	FB	Dr. Sarunas Peckus	SP
Bruce J. Benjamin	BJB	Ben Rapaport	BR
Ron A. Colter	RAC	Horst Reichert	HR
Gary L. Donachy	GLD	Dorothy E. Soares	DES
Alfred Dunhill Archive Collection	AD	Roy H. Webber	RHW
German Collector	GC	West Coast Collector	WC
		Arno Ziesnitz	AZ

Chapter IX. Celebration and Commemoration

To celebrate is to mark the occasion of a significant event, and to commemorate is to honor someone or something, two near-synonymous words. The photographs of the meerschaums selected for inclusion in this chapter immortalize and perpetuate important and memorable past events that someone selected as an appropriate remembrance. The event that stirred the emotion of triumph, defeat, joy, or sorrow must have been sufficiently historic and significant in its day, that a carver either was commissioned to create a representation in meerschaum, or the carver acted alone, impulsively, perhaps spontaneously, hoping that it would appeal to the buying public. The true origin of the majority of pieces illustrated herein may never be revealed, but there must have been an association of each meerschaum to an obvious antecedent; for some, the antecedent is patently obvious. In a few instances, I am not able to identify the exact relationship between the message and a specific event of the previous century, but I am confident that one probably exists.

Fierce battles and the peace pacts that followed, the courtship and marriage of both real and imagined people, the hunt as sporting game or man versus beast, and character scenes from opera and fiction are only a few of the subjects celebrated and commemorated in meerschaum. What appears in this chapter is, no doubt, a mere fraction of what must have been an endless array of subject matter. After all, the last century was filled with many momentous events, and I have to believe that Napoleon's defeat at Waterloo, Belgium, the sinking of the USS *Maine* in Havana Harbor, and everything in between, the warp and woof of that millennium and the previous century that became footnotes to history was probably considered worthy of preserving in meerschaum. The examples of celebration and commemoration found in this chapter are ample testimony of the meerschaum carver's passion to record these events.

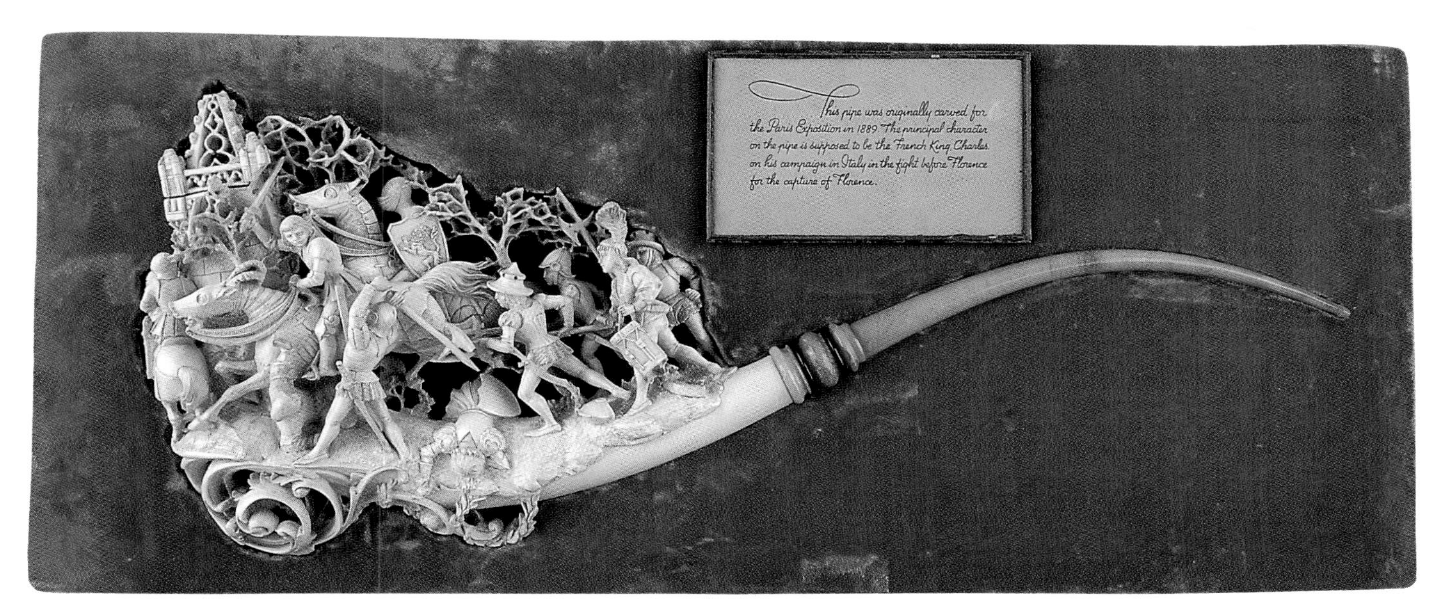

Allegedly made for the *Exposition Universelle,* Paris, 1889, this gigantic pipe relates an enlivened and frenetic battle scene; the uniforms and implements of war suggest the period as the Middle Ages, 28" l., 11" h., prob. French, ca. 1888. *Courtesy of the SP Collection.*

Closeup of the intricate detail surrounding the bowl and the reticulated meerschaum windcap in the shape of a tower.

This cheroot holder is a very unusual shape, and my guess is that it is the carver's interpretation of Admiral Horatio Nelson on the bridge of the H M S *Victory* at the Battle of Trafalgar on October 21, 1805, 7.5" l., 3" h., prob. English, ca. 1890. *Courtesy of the Author's Collection.*

Cigar holder in celebration of an Olympiad, a symbolic high-relief-carved runner, *Willi Olympier*, 6.5" l., 3.5" h., Emanuel Czapek, Prag, ca. 1890. (This piece may have been made for the Athens Games in 1896.) *Courtesy of the Author's Collection.*

Cheroot holder, figure of Britannia with shield and sword, the symbolic representation of the British Empire, after the ancient Roman name for the British Isles, 5.5" l., 3.5" h., prob. English, ca. 1890. *Courtesy of the Author's Collection.*

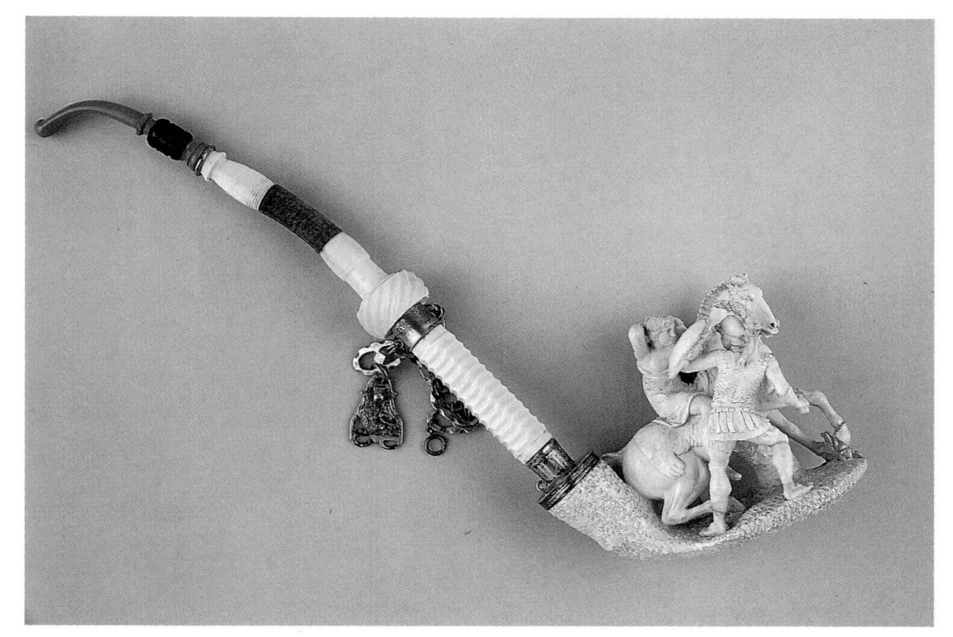

Considering this warrior's battle dress, the bowl may have commemorated in meerschaum the Nicolas Poussin painting, the *Rape of the Sabine Women*, the subjugation of people living northeast of Rome about 290 B. C., 5" l., 6" h., and a mixed media turned ivory and flexible hose push stem with amber mouthpiece, prob. Austrian-Hungarian, ca. 1875. (Alternatively, this may have been a rendition of the Greek hero, Theseus, doing battle with an Amazon.) *Courtesy of the SP Collection.*

Close-up of the combatants. *Courtesy of the SP Collection.*

Pipe bowl, Roman warrior seated in a chariot drawn by two lions, appears to be making unwanted advances, or it may be a different interpretation of the *Rape of the Sabine Women*. The silver-gilt band at the waist of both figures is the rim of the hinged windcap, 8" l., 8" h., prob. French, ca. 1880. *Courtesy of a WC Collector.*

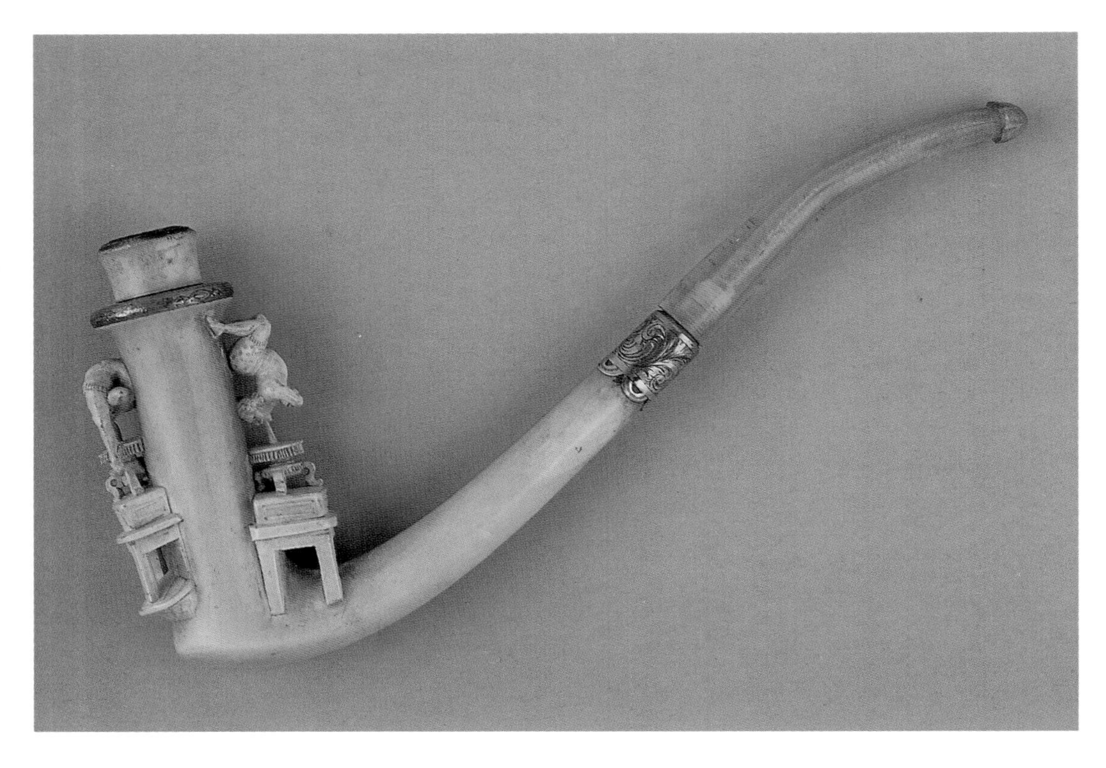

Cheroot holder celebrating the circus, a rendering of acrobats juxtaposed on either side of the bowl, 8.5" l., 4" h., prob. English, ca. 1890. *Courtesy of the AZ Collection.*

The (in-)famous Brigand pipe, described in Chapter III, a highwayman exulting to his family after a successful raid of looting and plundering, 18" l., 6" h., prob. Austrian-Hungarian, ca. 1875. *Courtesy of the SP Collection.*

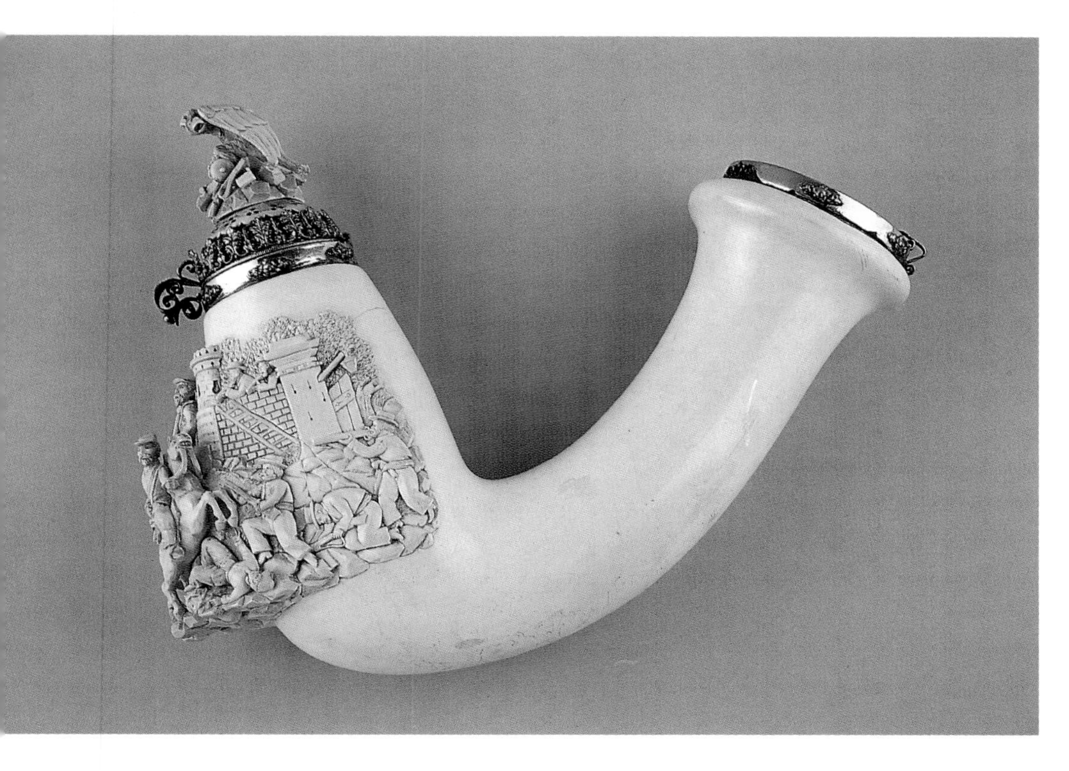

Pipe bowl, bas- and high-relief-carved American Civil War battle scene, silver windcap with meerschaum spread eagle finial, 13" l., 10" h., prob. American, ca. 1875. (From the Half and Half Collection.) *Courtesy of the SP Collection*.

Cheroot holder, exquisite and ornate expression celebrating the hunt, three mounted riders accompanied by nine hounds in hot pursuit of a stag, 12.5" l., 4.5" h., prob. German or Austrian-Hungarian, ca. 1875. *Courtesy of the FB Collection.*

Pipe, allegorical motif, a symbolical yet indecipherable narration, perhaps made for one of the world's fairs in celebration of the arts and industries; the escutcheon on the front is incised "E Viva Italia," 16" l., 4" h., Savinelli, Milano e. Genova, ca. 1900. *Courtesy of the SP Collection.*

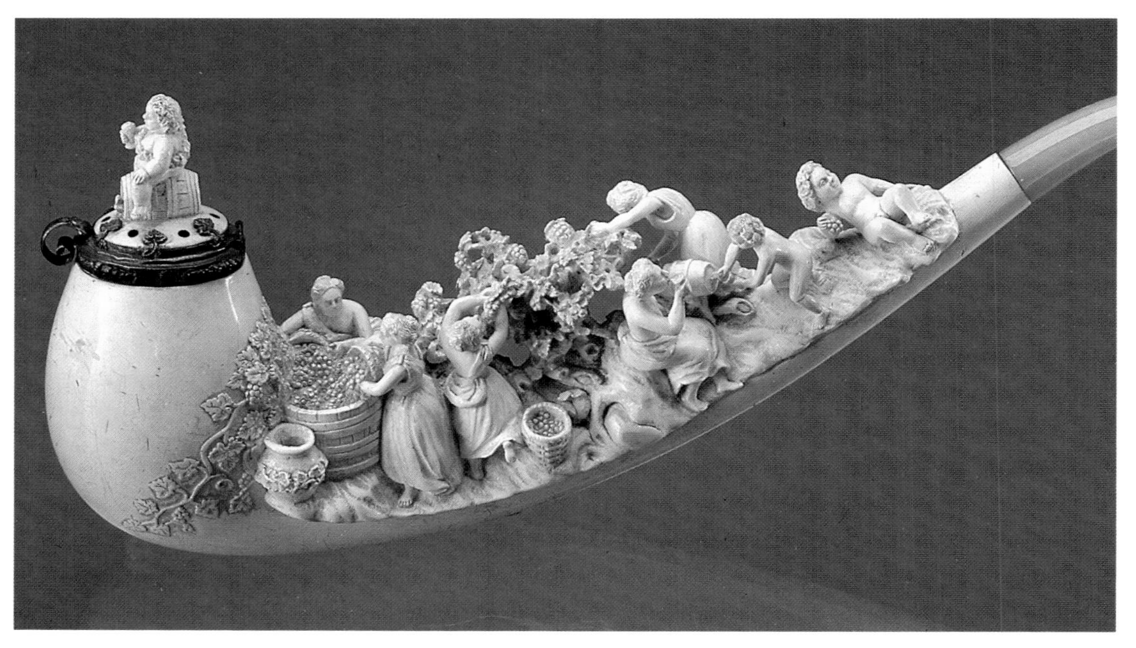

Pipe, women and children celebrating the harvest of the grape, bacchanalian-style, 14.5" l., 5" h., Austrian-Hungarian, ca. 1875. *Courtesy of the GC Collection.*

Pipe, a vastly different expression of the Three Graces rising above the shank, silver and amber windcap, 10.75" l., 4.25" h., prob. German, ca. 1880. *Courtesy of the AZ Collection.*

Celebrating the fruits of the sea, this pipe's motif portrays a woman astride a conch, with crayfish, mussels, and other assorted crustaceans, 11.5" l., 5" h., prob. American, ca. 1875. *Courtesy of the Author's Collection.*

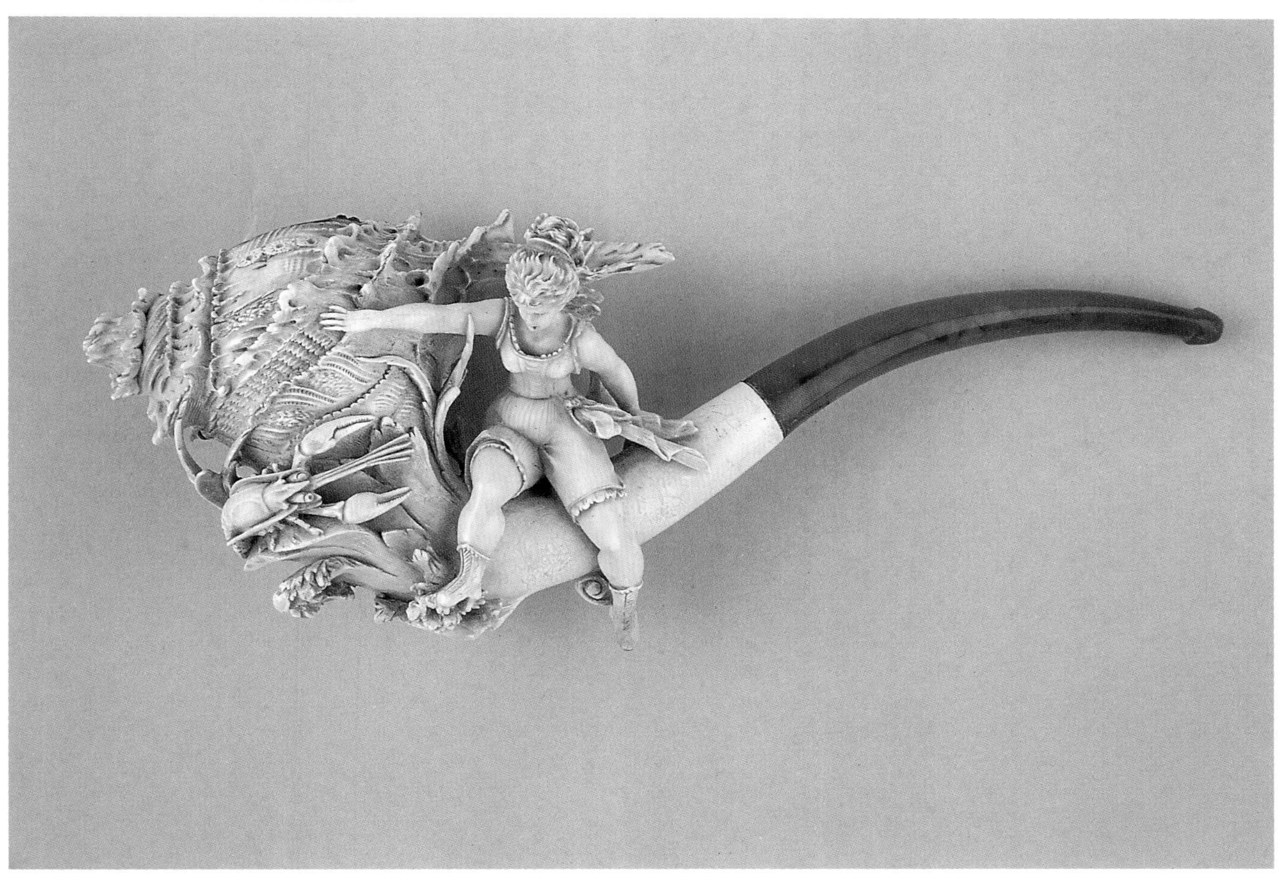

Chapter X. Femmes Fatales

The femme fatale, the irresistibly attractive woman, the gentler sex, the siren, the vixen, was an ever-popular motif. The beauty, grace, comeliness, and the allure of the female, dressed and undressed, has always been a major area of interest for all artists, and the meerschaum carver was no exception. The meerschaums in this domain can be classified into two groups, figurative heads and full figures. In the former group, hundreds of figurative expressions were crafted, a range that included those wearing elegant, broad-brimmed hats, others exhibiting complex hairstyles—ringlets and intricately-braided chignons—and yet others using pince-nez, or carrying a portmanteaux or parasol.

But the best of class are those meerschaums that offer a view of at least the torso, and better, the full physique. These, too, were made in large quantity, and the variety of motifs was near-infinite: the recumbent and the upright; alone, in groups, and accompanied by pets; the fully clothed, partially clothed, and unclothed; and the innocent, the beguiling, the narcissistic, and even the lady of the night. Less frequently, the young girl or the nymphet was captured in meerschaum, but in nowhere near the volume as the mature and sophisticated adult female. Some of the best examples in this latter category were selected to illustrate the variety and assortment of all the possible subject matter in this domain.

Cheroot holder, seated nude playing lyre, 6" l., 2.5" h., G. A. F., American, ca. 1880. *Courtesy of the Author's Collection.*

Cheroot holder, Gay Nineties-dressed woman astride a hand holding a champagne flute, 4" l., 2.5" l., prob. French, ca. 1880. *Courtesy of the Author's Collection.*

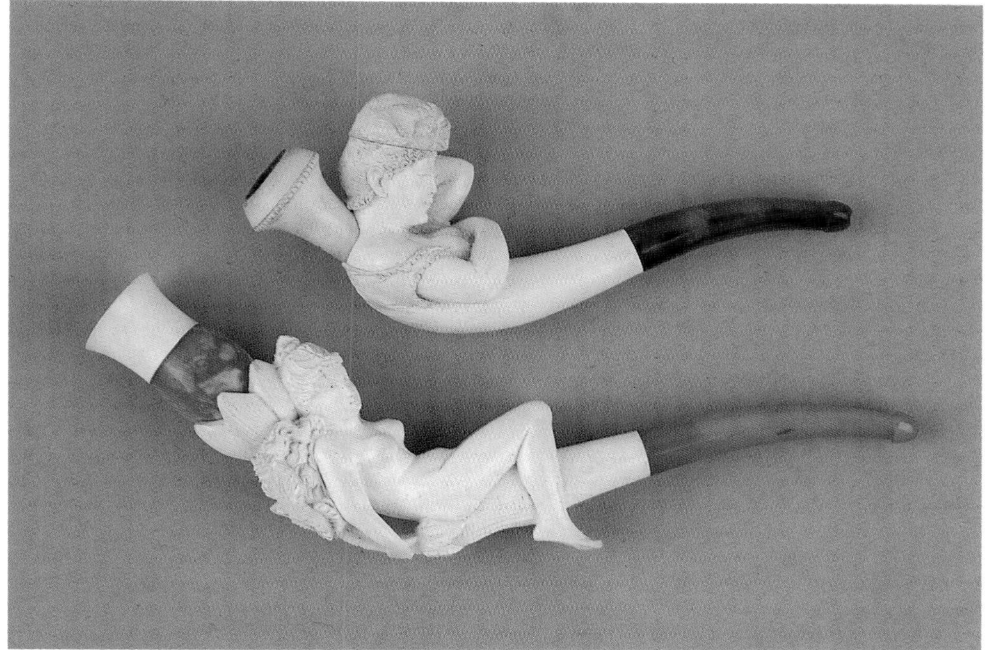

Cheroot holders, reclining woman wearing a bandanna, 4.5" l., 2" h., prob. German, ca. 1880 (top); nude, 6" l., Gambarini, Napoli, ca. 1880 (bottom). *Courtesy of the Author's Collection.*

Cheroot holder, woman seated on an ottoman, 6" l., 3" h., C.W. Möller, Berlin, ca. 1875. *Courtesy of the Author's Collection.*

Cheroot holder, two young maidens captivated by a swan, 7.25" l., 4.5" h., prob. German, ca. 1880. *Courtesy of the SP Collection.*

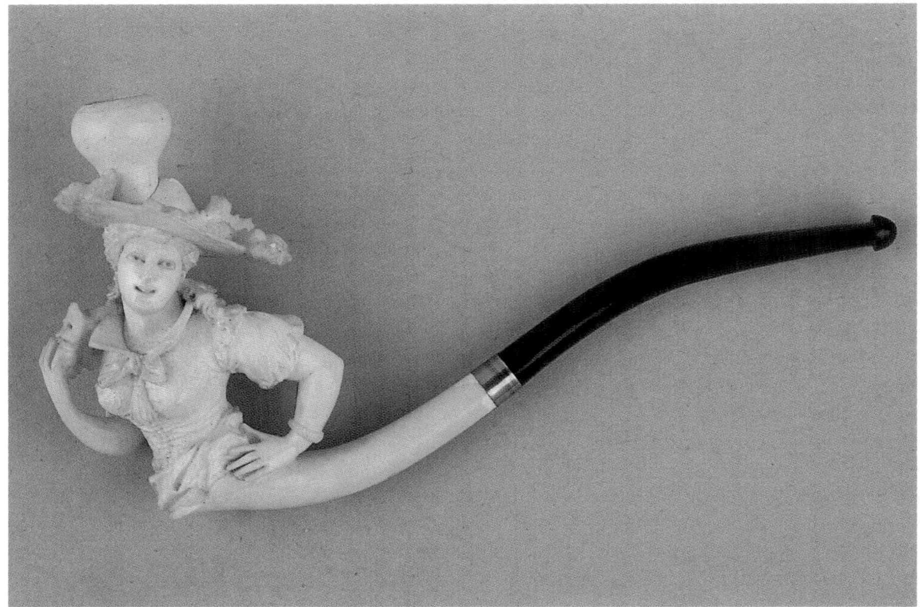

Cheroot holder, bust of buxom woman in large hat, 6" l., 3" h., prob. French, ca. 1880. *Courtesy of the Author's Collection.*

Cheroot holder, bust of a young lady holding a fan, accented with amber coloring bowl, 6.5" l., 3.5" h., C.C., Paris, ca. 1880. *Courtesy of the Author's Collection.*

Cheroot holder, woman observing bird on her left shoulder, 6" l., 3.5" h., Emanuel Czapek, Prag, ca. 1880. *Courtesy of the Author's Collection.*

The first of four similar configurations, cheroot holder of seated nude holding a birdcage, 7" l., 6" h., prob. Austrian, ca. 1880. *Courtesy of the SP Collection.*

Cheroot holder, seated girl knitting, 7" l., 4.5" h., Ludwig Hartmann & Eidam, Wien, ca. 1880. *Courtesy of the SP Collection.*

Cheroot holder, girl knitting and reading a book which is resting on her left knee, 5.5" l., 3.5" h., prob. Austrian-Hungarian, ca. 1880. *Courtesy of the FB Collection.*

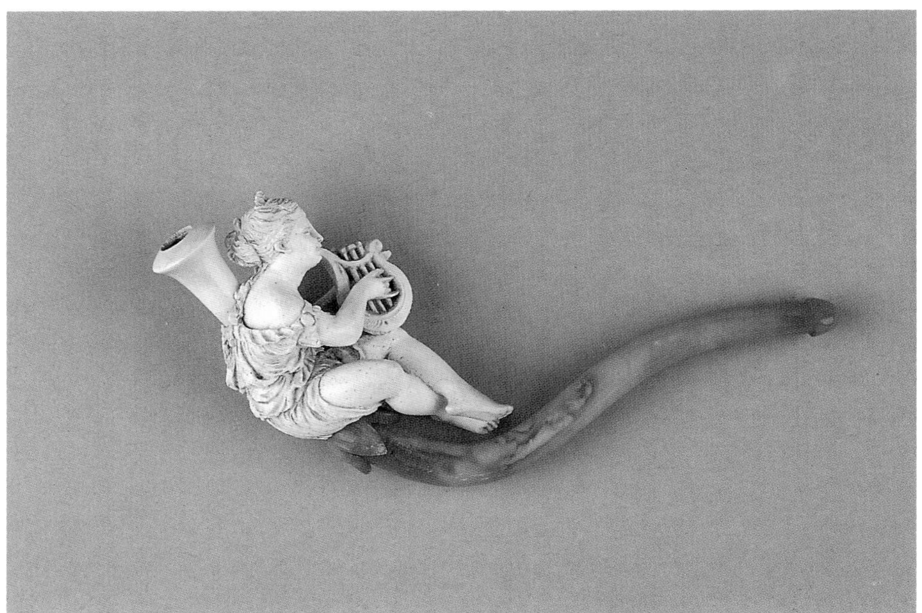

Cheroot holder, seated woman in Roman vestment playing lyre, 4.5" l., 2.5" h., Edoardo Flegel, Milano, ca. 1880. *Courtesy of the FB Collection.*

Cheroot holder, young boy cavorting with young girl, 6.75" l., 3" h., prob. German, ca. 1880. *Courtesy of the AZ Collection.*

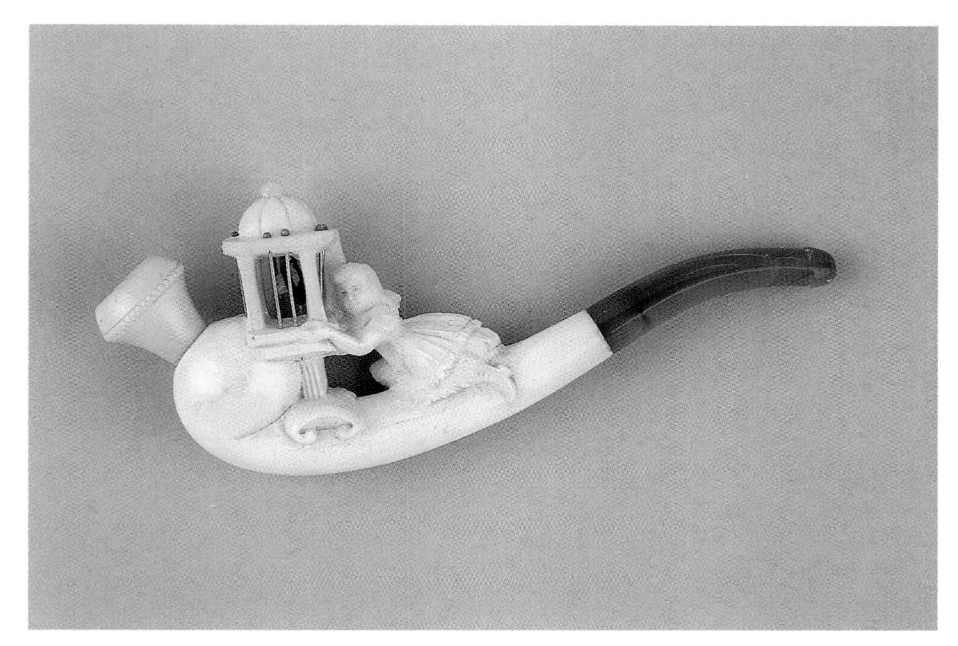

Cheroot holder, young girl looking at birdcage, 4.5" l., 2" h., prob. English or American, ca. 1890. *Courtesy of the Author's Collection.*

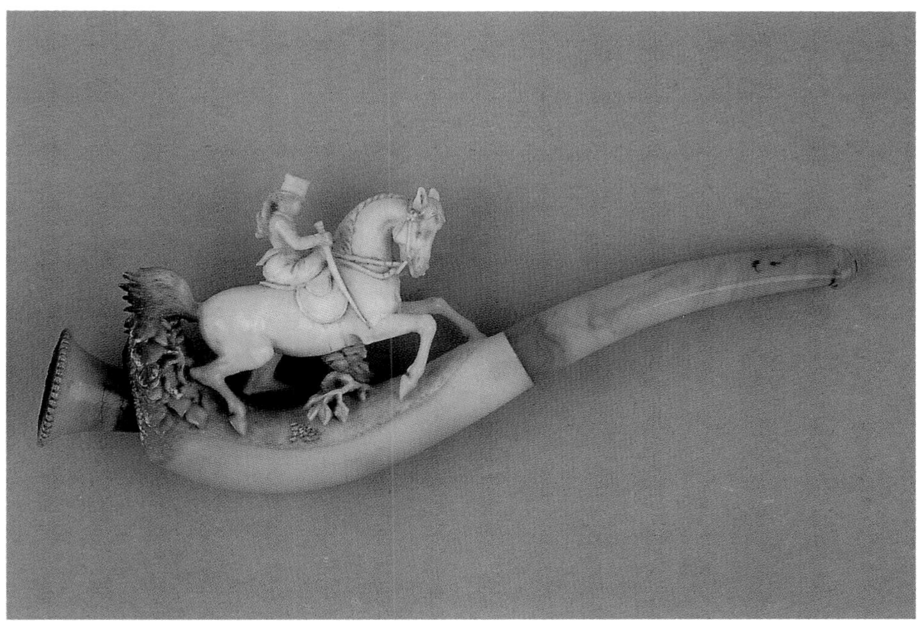

Cheroot holder, woman in long skirt and top hat riding side-saddle, 6.5" l., 2.5" h., prob. German, ca. 1880. *Courtesy of the AZ Collection.*

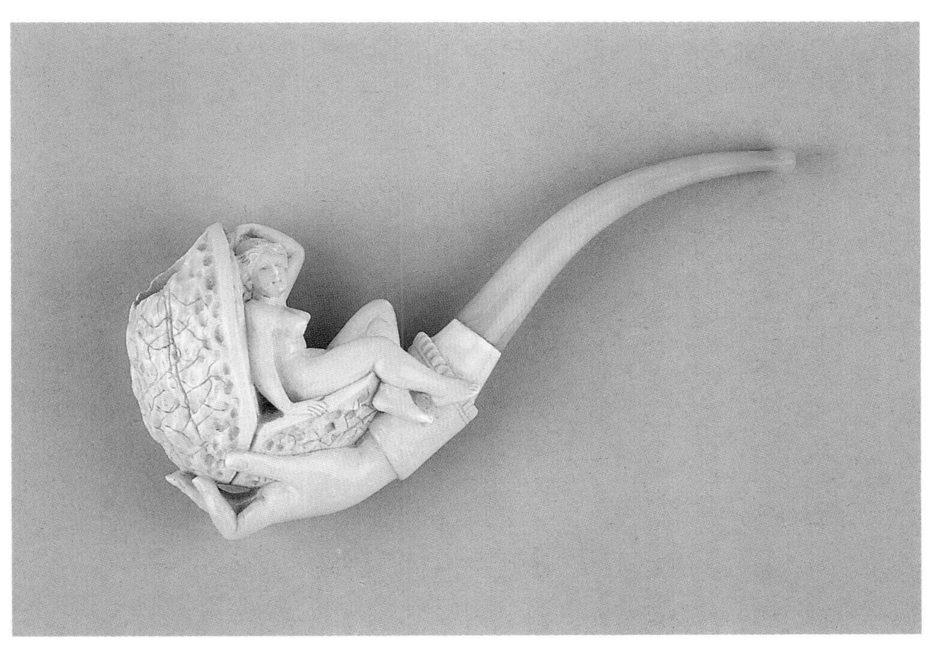

Pipe, hand holding open walnut shell within which is a seated nude, 6.5" l., 3" h., prob. American, ca. 1890. *Courtesy of the RC Collection.*

Cheroot holder, bacchante, 7.5" l., 3" h., prob. German, ca. 1880. *Courtesy of the GC Collection.*

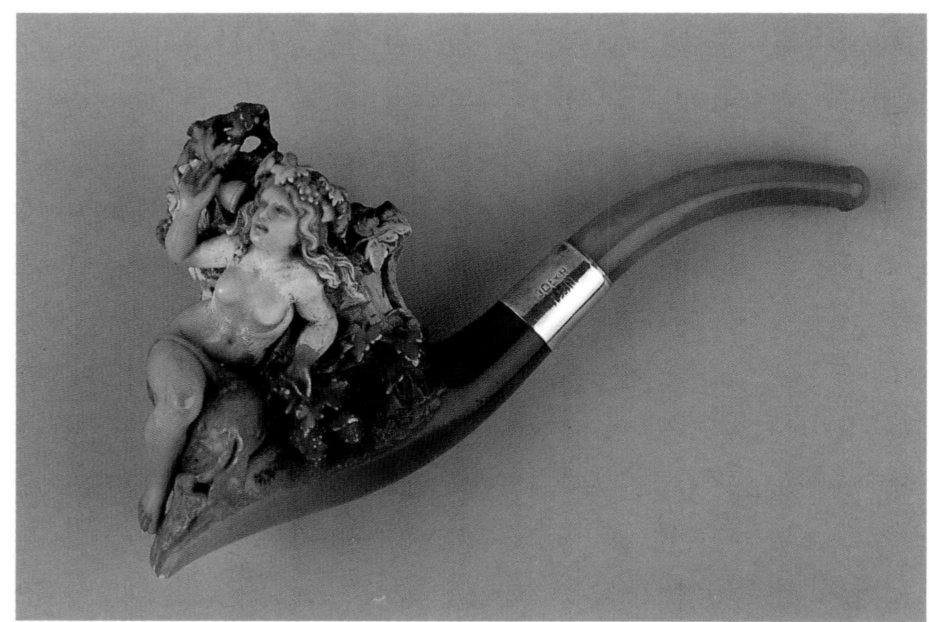

Pipe, bacchante, 6.75" l., 3.5" h., prob. German or Austrian-Hungarian, cased stamped "*Dem Fortschritte*, 1873 Exhibition, Vienna." *Courtesy of the AZ Collection*.

Cheroot holder bowl, young girl with port-monnaie and grapes, 2.5" l., 3.25" h., prob. German, ca. 1875. *Courtesy of the AZ Collection*

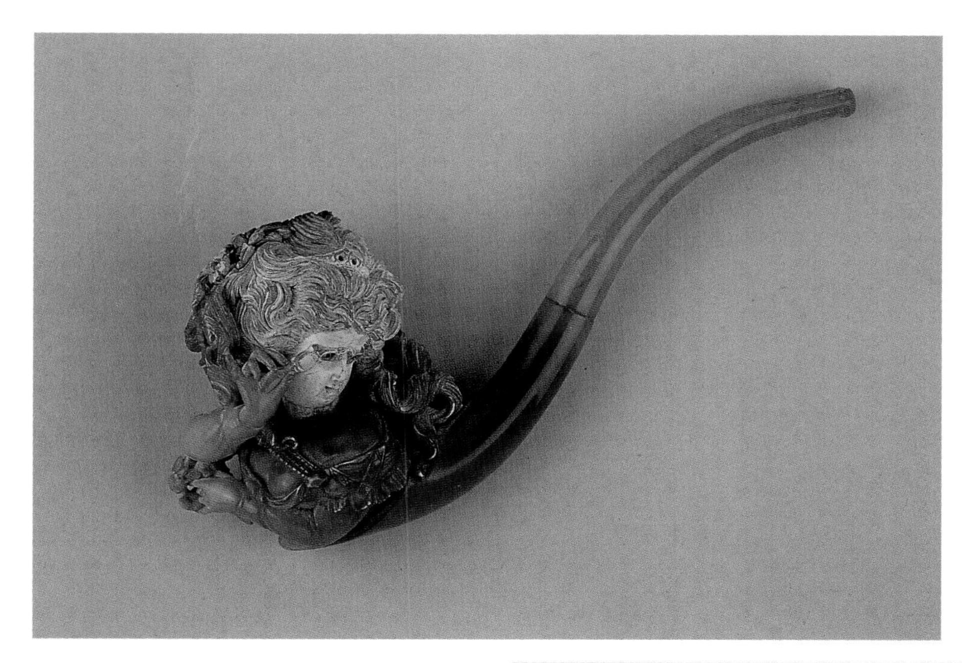

Pipe, bust of lady, inset glass eyes and lorgnette, 8" l., 3.5" h., prob. French, ca. 1880. *Courtesy of the Author's Collection.*

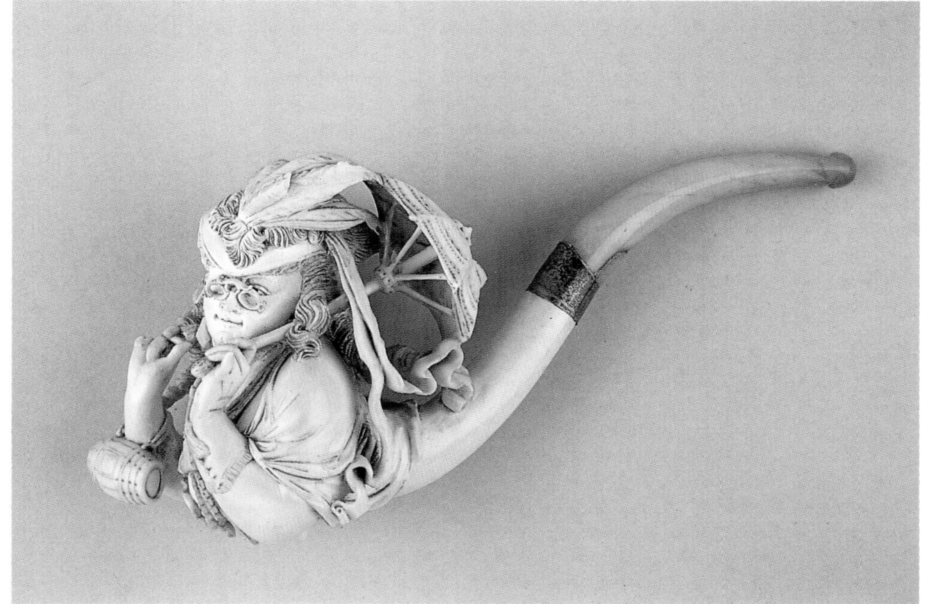

Pipe, bust of haute couture-dressed Victorian woman in ornate plumed hat, inset glass eyes and pince-nez, carrying parasol and portmonnaie, 9.5" l., 5.5" h., prob. English, ca. 1875. *Courtesy of the SP Collection.*

Pipe bowl, bust of haute couture-dressed Victorian woman, beribboned hair, inset glass eyes and lorgnette, holding fan, 6.5" l., 6.5" h., attributed to Gustav Fischer of Boston, ca. 1875. *Courtesy of the SP Collection.*

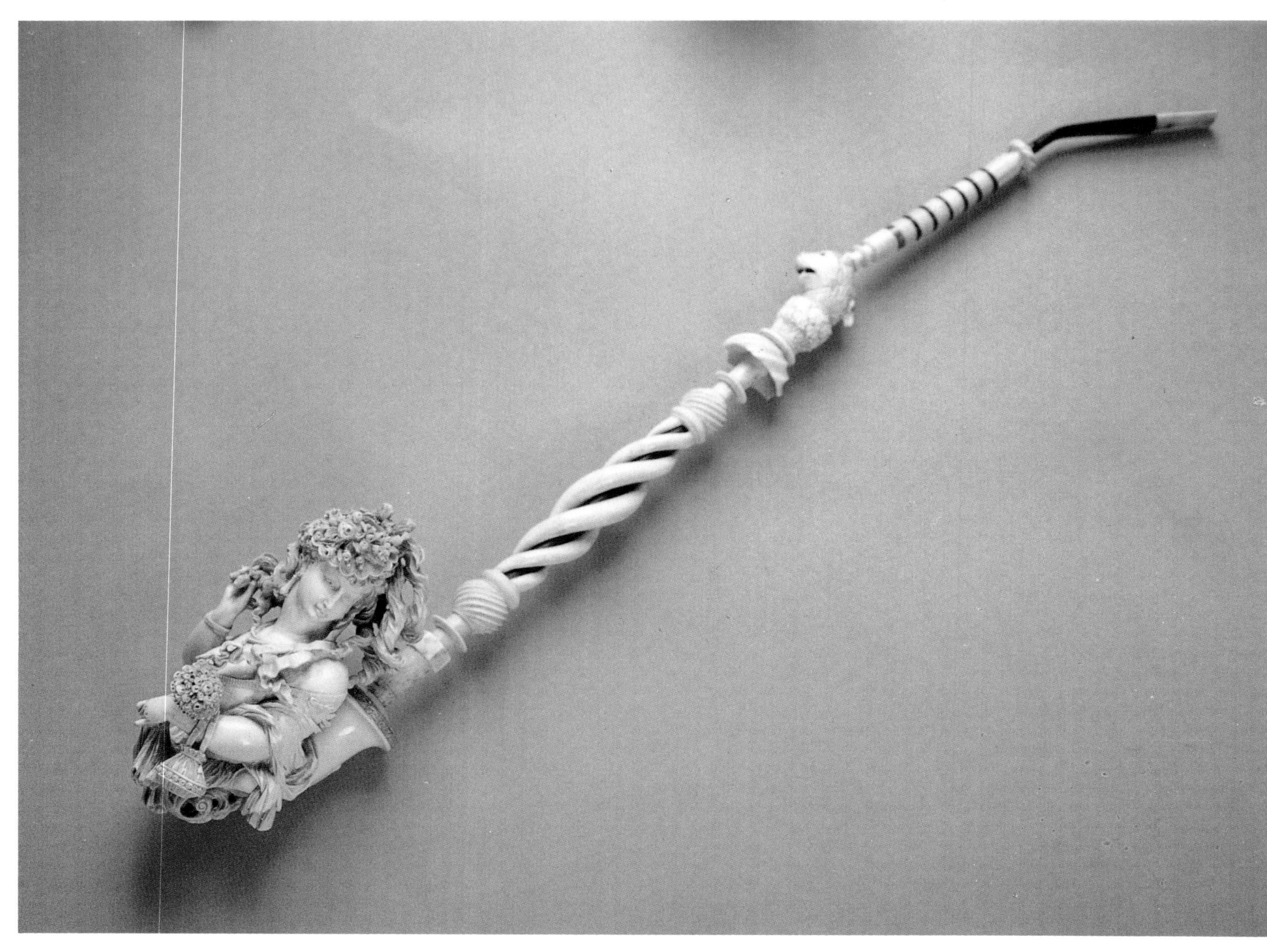

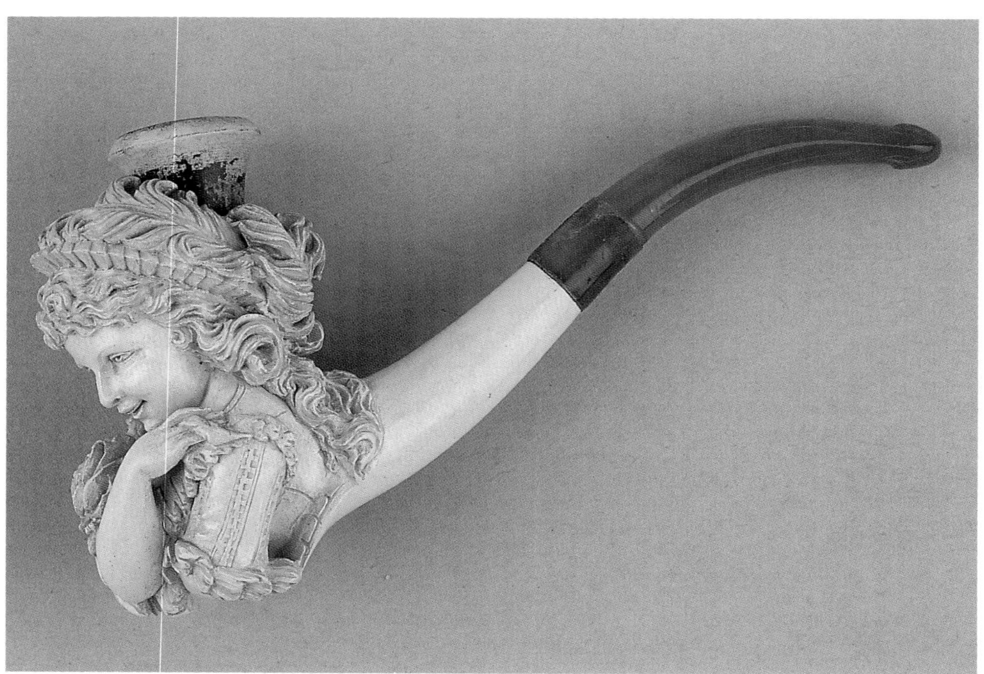

Pipe, bust of haute couture-dressed Victorian woman in floral hat, carrying port-monnaie and wrist corsage, 7" l., 5.5" h., push stem of turned ivory with ebony band accents, inset figurative head of ornamental leopard with ruby eyes, amber mouthpiece, 28.5" l. overall, prob. English or French, ca. 1875. *Courtesy of the JTB Collection.*

Cheroot holder, bust of nymphet clasping rosebud, 5.5" l., 2.5" h., F. J. Kaldenberg, New York, ca. 1880. *Courtesy of the Author's Collection.*

Cheroot holder, head and leg of woman, shoe terminated as mouthpiece, 4" l., prob. American or English, ca. 1885. *Courtesy of the SP Collection.*

Cheroot holder set, recumbent woman, composed of two holders, one of the upper torso, the other, the leg terminated as mouthpiece, 5.5" l., prob. French, ca. 1885. *Courtesy of the SP Collection.*

Cigar holder, incised carving of dancing man and woman, 6" l., prob. German, ca. 1875. *Courtesy of the SP Collection.*

Cheroot holder, torso of recumbent woman, chignon and braided hairdo, 8.5" l., 2.5" h., prob. English, ca. 1885. *Courtesy of the JTB Collection.*

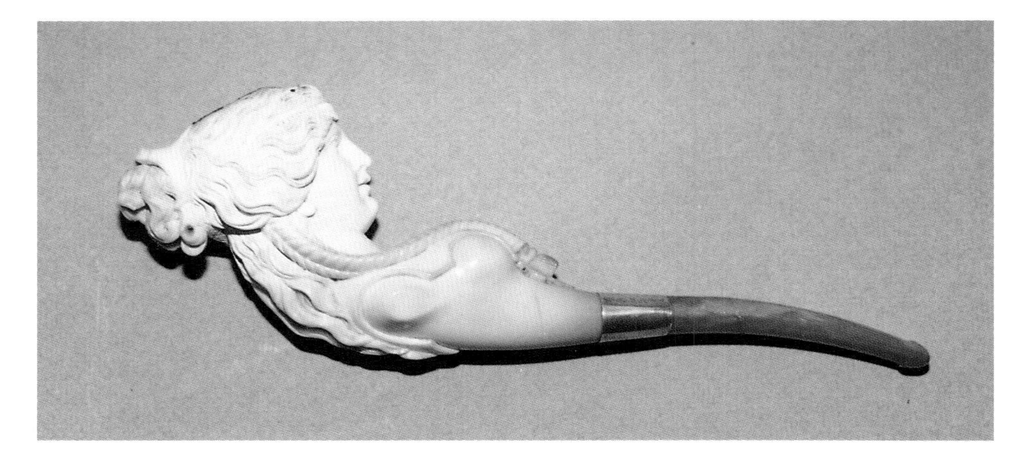

Pipe, woman feeding cockatoo, 6.5" l., 2.75" h., prob. English, ca. 1875. *Courtesy of the JTB Collection.*

Cheroot holder, reclining woman holding bouquet of flowers, one leg terminated as mouthpiece, 7" l., 3" h., prob. English, ca. 1875. *Courtesy of the JTB Collection.*

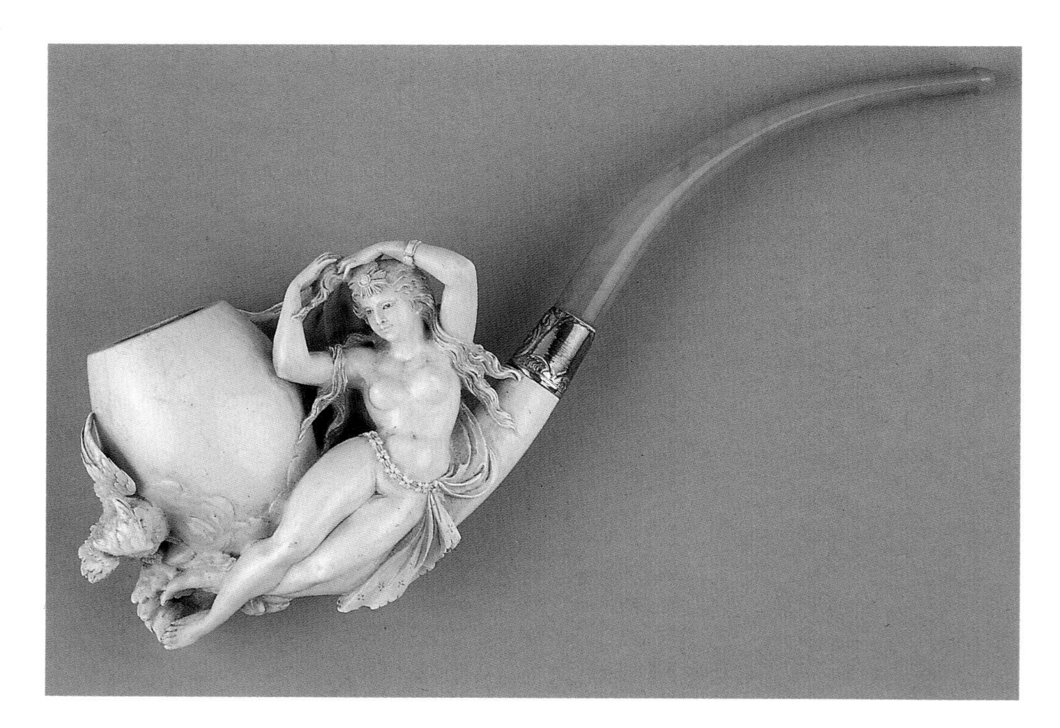

Pipe, reclining nude under observation of owl, 10.75" l., 5" h., prob. French, ca. 1875. *Courtesy of the SP Collection.*

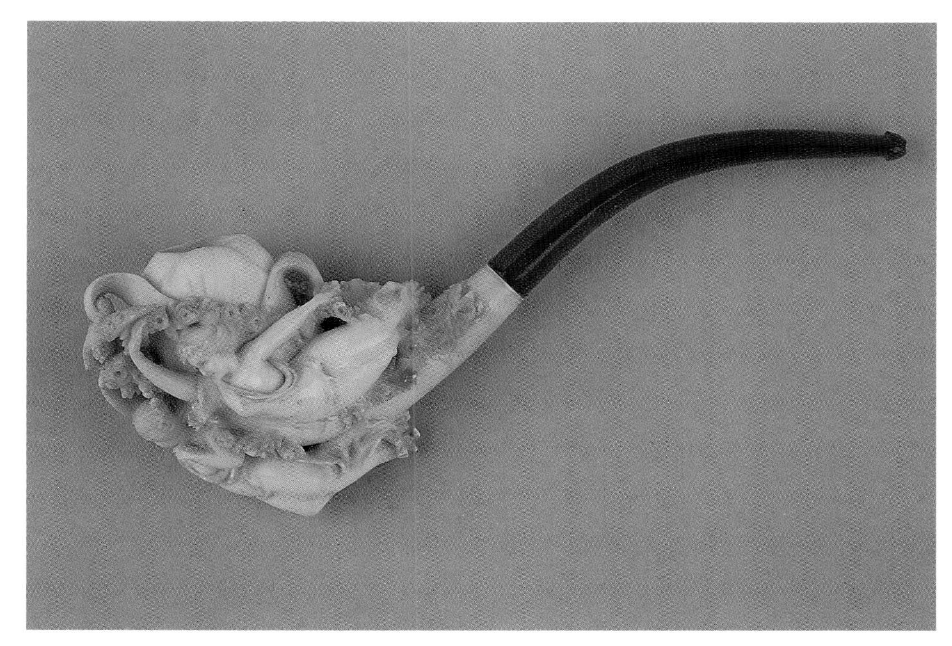

Pipe, two nudes, one astride, one in front of tulip-shaped bowl, 8.5" l., 3.5" h., prob. American, ca. 1875. *Courtesy of the Author's Collection.*

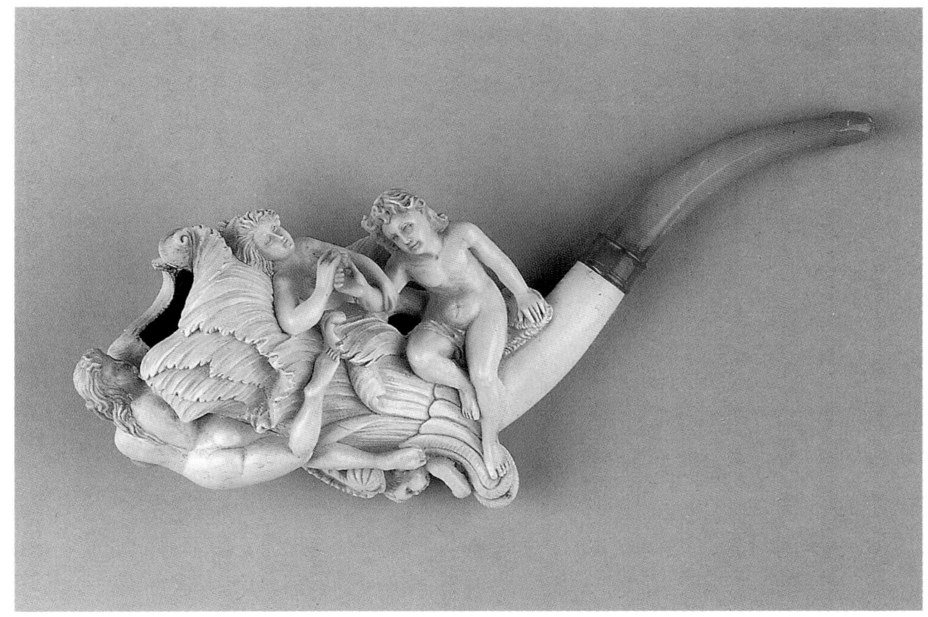

Pipe, three consorting nudes and putto, 8.5" l., 5" h., attributed to Gustav Fischer, Boston, ca. 1895, exhibited in *Boston Artifacts, Objects Collected for Where's Boston?,* 1975, 45. *Courtesy of the FB Collection.*

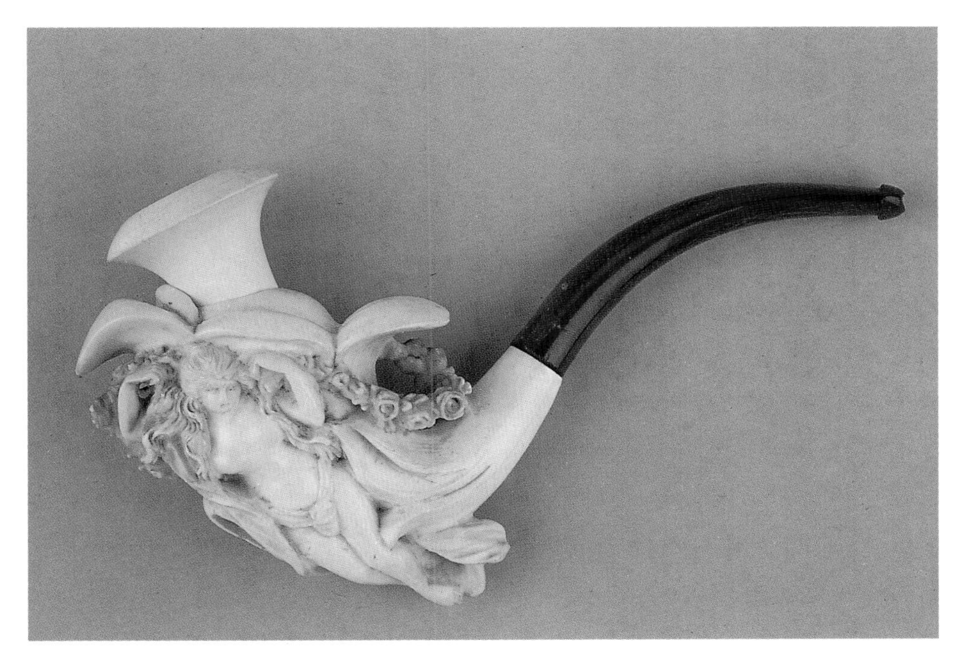

Pipe, nude surrounded by palm fronds and garlands, 7" l., 4.5" h., C.P.F., American, ca. 1875. *Courtesy of the RC Collection.*

Pipe, pre-colored standing semi-nude astride bowl, another semi-nude beneath bowl, and small bird perched on the shank, 9" l., 4.75" h., prob. American, ca. 1875. *Courtesy of the FB Collection.*

Cheroot holder, standing nude with garland necklace, accompanied by cherub, both guarded by fierce-looking mythical animal, 9" l., 5.25" h., prob. Austrian-Hungarian, ca. 1875. *Courtesy of the GC Collection.*

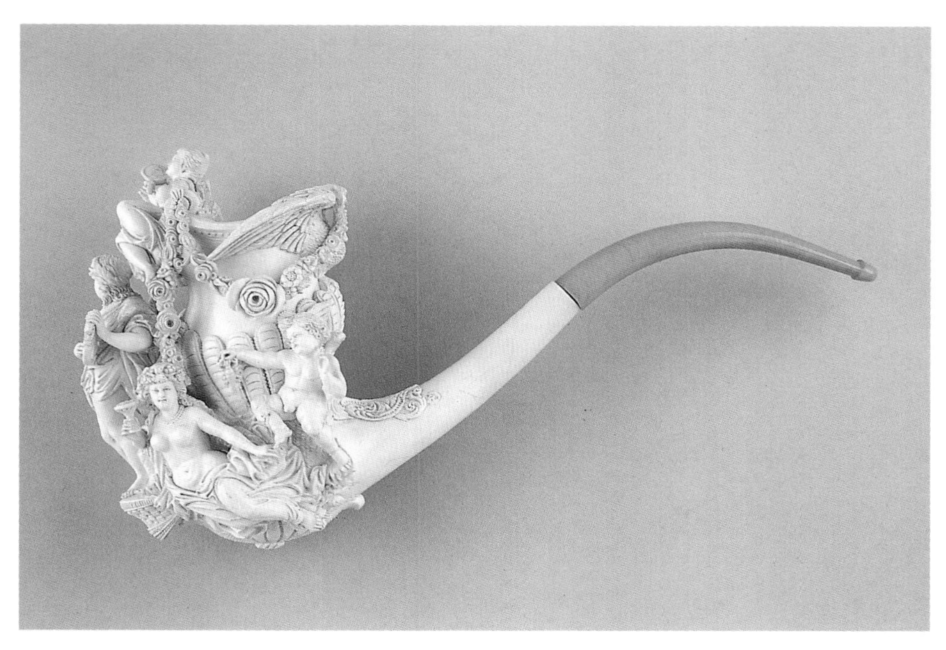

Pipe, obverse of an in-the-round scene of frolicking maidens, one of which poses narcissistically in front of a mirror, cherub and eagle accompany, 13.5" l., 7" h., case stamped *"Dem Fortschritte"* (*Weltaustellung, Wien,* 1873). *Courtesy of the SP Collection.*

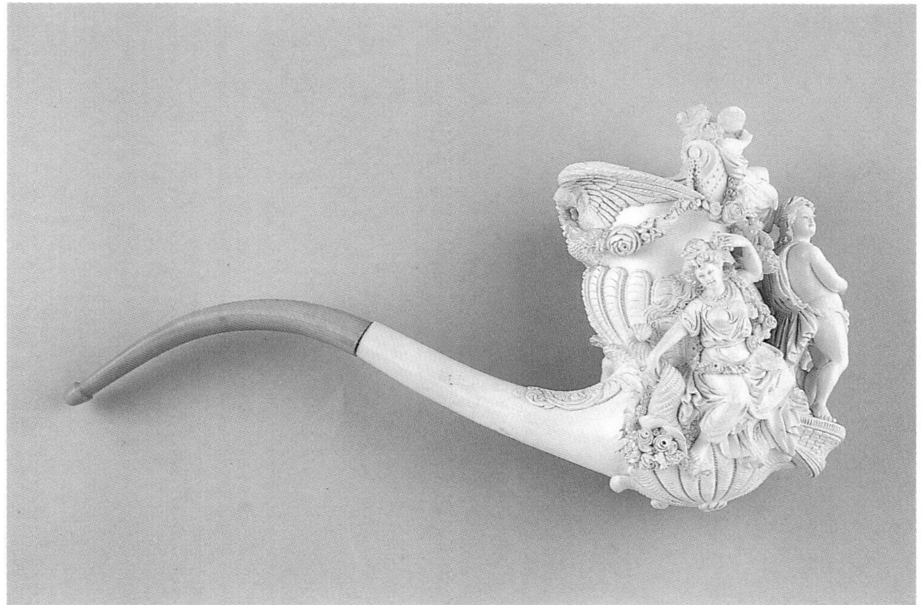

Reverse of this pipe.

Pipe, Venus (Italian goddess of gardens and spring) or Aphrodite (Roman goddess of love and beauty), seated between two cherubim, the one on her right holding reins of two dolphins, the other playing lyre, 10" l., 5.5" h., prob. French, ca. 1875. *Courtesy of the FB Collection.*

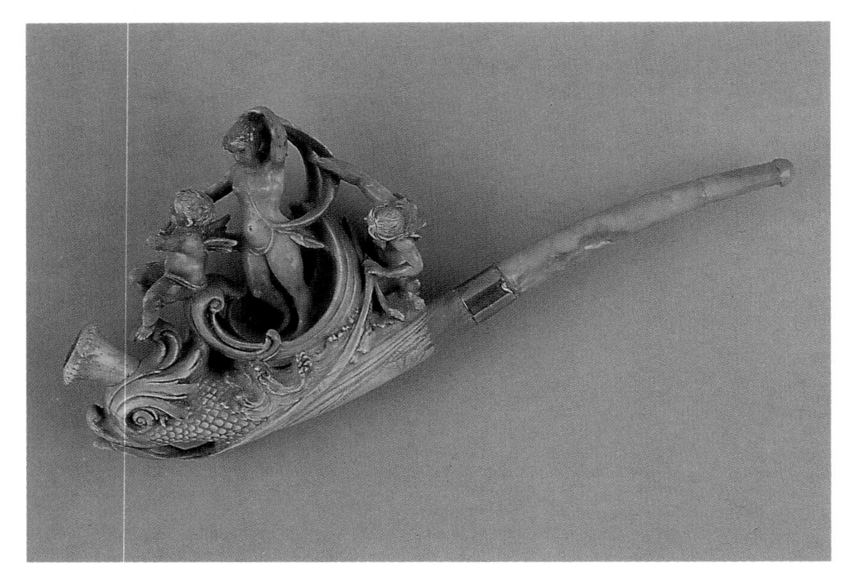

Cheroot holder, nereid and cherubs riding atop stylized dolphin, 8.5" l., 3.5" h., Alexandre, St. Petersburg, ca. 1880. *Courtesy of the SP Collection.*

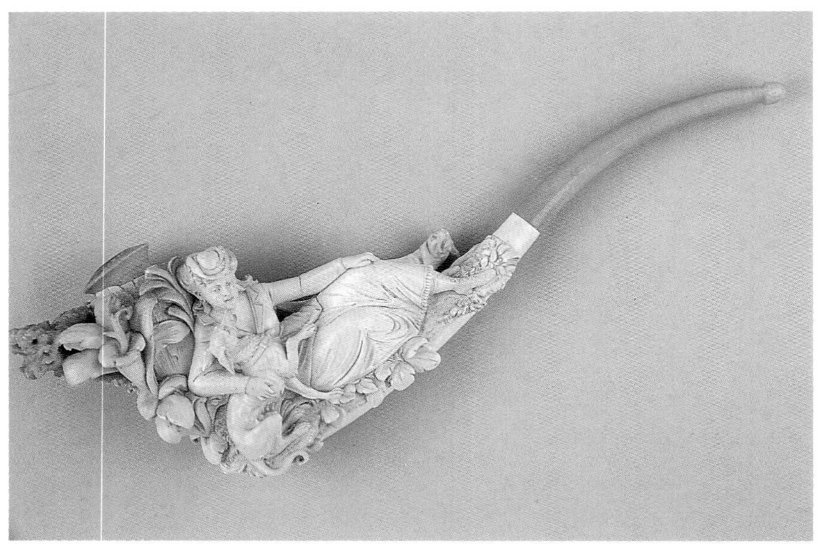

Pipe, woman reclining against tree trunk, accompanied by dog, 12.5" l., 4.5" h., prob. English, ca. 1875. *Courtesy of the SP Collection.*

Pipe, Empress Elizabeth of Austria wearing elaborate flounced dress and holding bouquet, seated on wide-backed chair that forms the bowl, 14" l., 6" h. prob. Austrian Empire (allegedly exhibited at the Crystal Palace Exhibition, 1851). *Courtesy of the GC Collection.*

Pipe, scantily clad woman seated alongside birds, dog, fruit, flowers, and birdcage, 10" l., 4.5" h., prob. English, ca. 1875. *Courtesy of the AZ Collection.*

Pipe, standing nude, bas-relief-carved ram's head on front of bowl, coloring bowl, 8.5" l., 5" h., prob. American, ca. 1875. *Courtesy of the Author's Collection.*

Below:
Pipe, three nudes at bath, satyr lurking at lower right, 10" l., 5" h., prob. American or English, ca. 1880. *Courtesy of the MG Collection.*

Pipe, art nouveau style hookah, semi-nude on meerschaum-accented amber sphere and ebony pedestal, 12.5" h., prob. French, ca. 1900. *Courtesy of a WC Collector.*

Pipe, woman playing piano, pipe organ to the rear, 9" l., 3" h., prob. German, ca. 1890. *Courtesy of the FB Collection.*

Bowl, figurative head of female Nubian with bi-color accents on earrings and teeth, 5" l., 3.5" h., accompanied by albatross bone push stem with horn mouthpiece, William Astley's, London, ca. 1885. *Courtesy of the Author's Collection.*

Pipe, figurative head of nun (religious subject matter is relatively rare), 6" l., 3" h., D. P. Ehrlich Company, Boston (most likely carved by Gustav Fischer Sr.), ca. 1890. *Courtesy of the Author's Collection.*

Chapter XI. Of Mice and Men

Following the chapter on the femme fatale, man, and mouse, as well as all the others in the animal kingdom, deserve their own chapter. In reality, although the title of this chapter takes liberties with that of a John Steinbeck novel, the content of this chapter is about Man and all manner of animal, fish, and fowl. Just as popular as the female figurative, the male counterpart was also produced in substantial quantities, as figurative heads—potentates, pashas, poets, politicians, patriots, and many others—and as in-the-round scenes of countless masculine themes that defy categorization. As well, the animal was a dominant force in meerschaum, and a large quantity of feral and ferocious animals, such as the cougar, lion, and wolf, was produced.

The several male, animal, and other assorted motifs illustrated in this chapter are just representative of the broad array of masculine themes portrayed on pipes and holders conceived during the period under study. I can only surmise that many more than the selected meerschaums pictured here are sequestered in private collections unknown to me.

Pipe, four figurative male heads rise above the shank, silver windcap, 10" l., 1.5" h., prob. French, ca. 1900. *Courtesy of the SP Collection.*

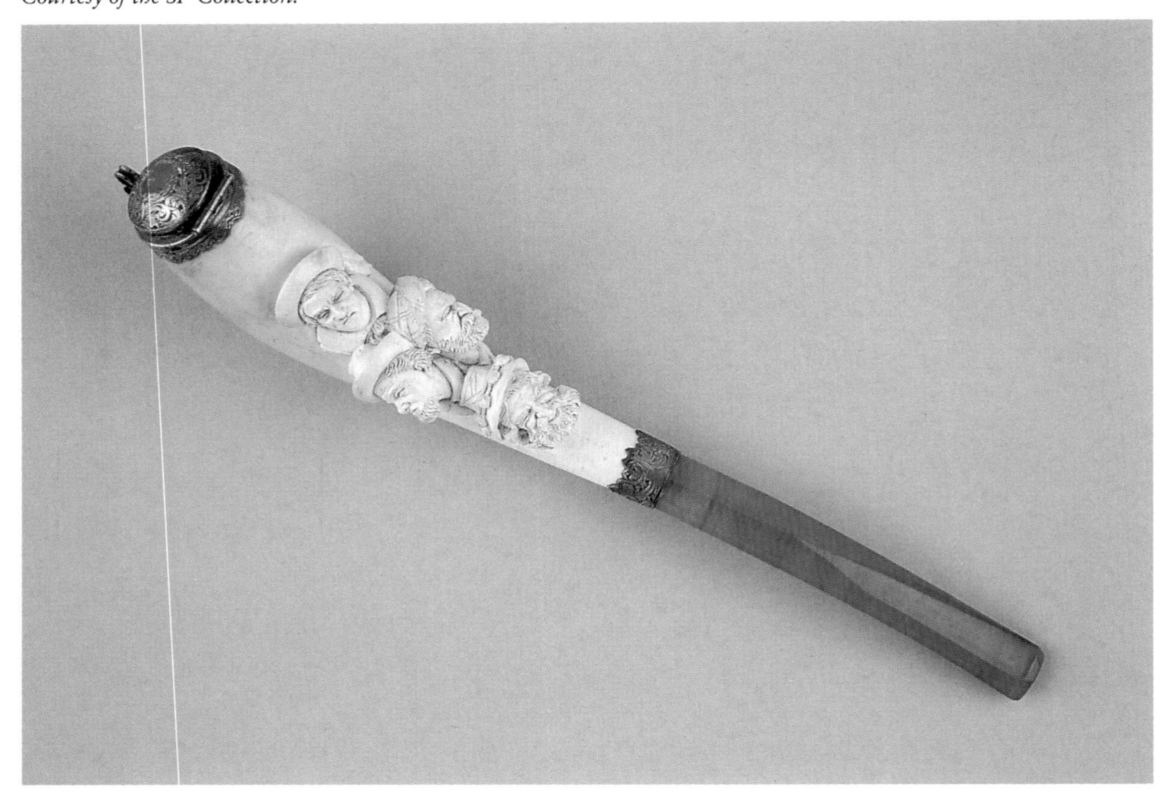

Cheroot holder, tradesman's motif, bust of carpenter holding wood plane, 5" l., 3" h., C. Schnally, Bremen, ca. 1885. *Courtesy of the Author's Collection.*

Cheroot holder, figurative head of bearded gentleman, 5.25" l., 2" h., G. Fischer, Boston, ca. 1890. *Courtesy of the Author's Collection.*

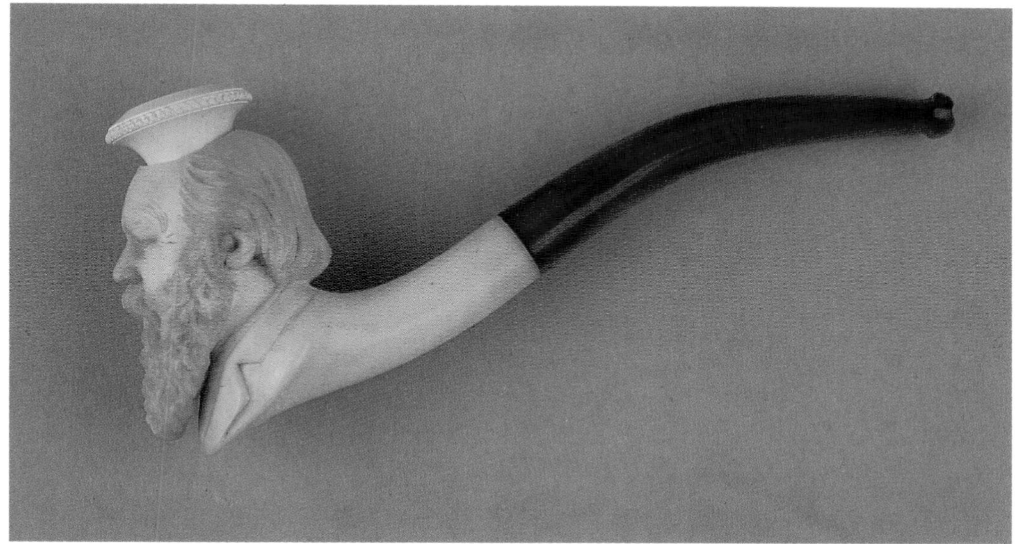

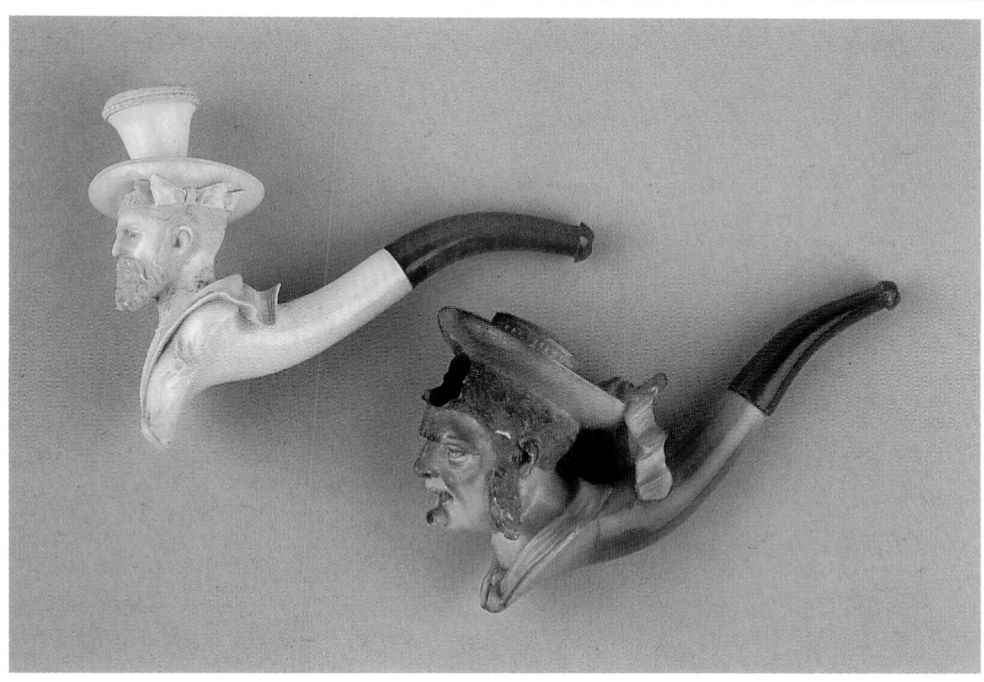

Cheroot holders, figurative head of English sailor, hat is incised "H.M.S. *Tartar*," 4" l., 2.5" h., C.W. Möller, Berlin, ca. 1890 (top); figurative head of English sailor smoking clay cutty, 4.5" l., 2.5" h., Emanuel Czapek, Prag, ca. 1890 (bottom). *Courtesy of the Author's Collection.*

Cheroot holder, figurative head of Arab with the 'mother of all' beards, 4" l., 4.5" h., prob. French, ca. 1885. *Courtesy of the RC Collection.*

Cheroot holder, figurative head of Nubian male with earrings, 5.5" l., 2.75" h., prob. American, ca. 1890. *Courtesy of the SP Collection.*

Cheroot holder, figurative head of a Paddy, 4" l., 1.75" h., prob. English, ca. 1890. *Courtesy of the Author's Collection.*

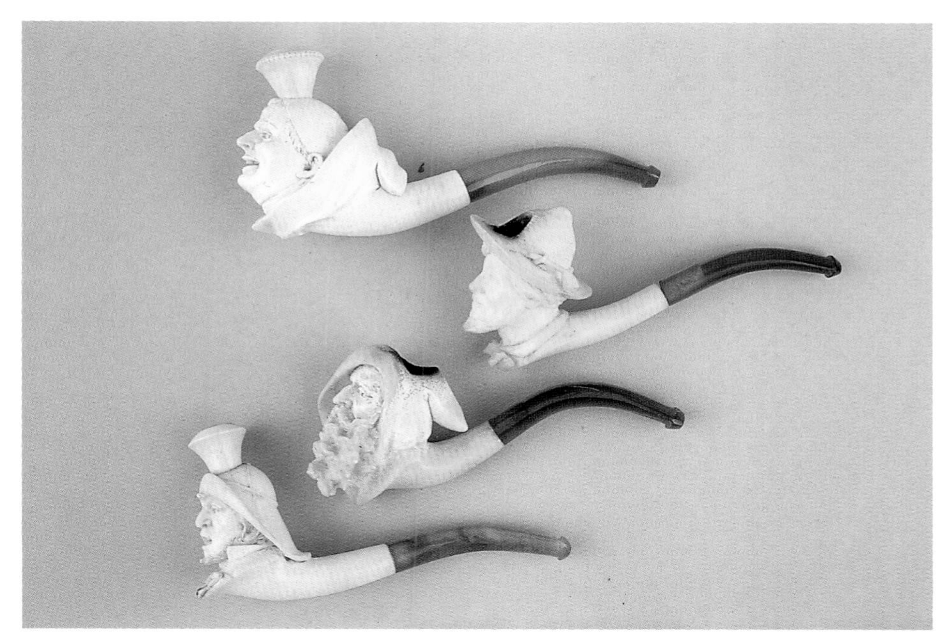

Cheroot holders, figurative head of smiling friar in cowl, 5.5" l., 2.5" h., G. A. F., American, ca. 1890 (top); figurative head of man in western-style hat, accent of clay pipe in brim, 5" l., 2" h., BBB, English, ca. 1890 (center top); figurative head symbolic of the North Wind, 5" l., 2" h., G. A. F., American, ca. 1890 (center bottom); figurative head of Sou'easter, symbol of the New England fisherman, 5" l., 2.75" h., Made in Vienna, ca. 1890 (bottom). *Courtesy of the Author's Collection.*

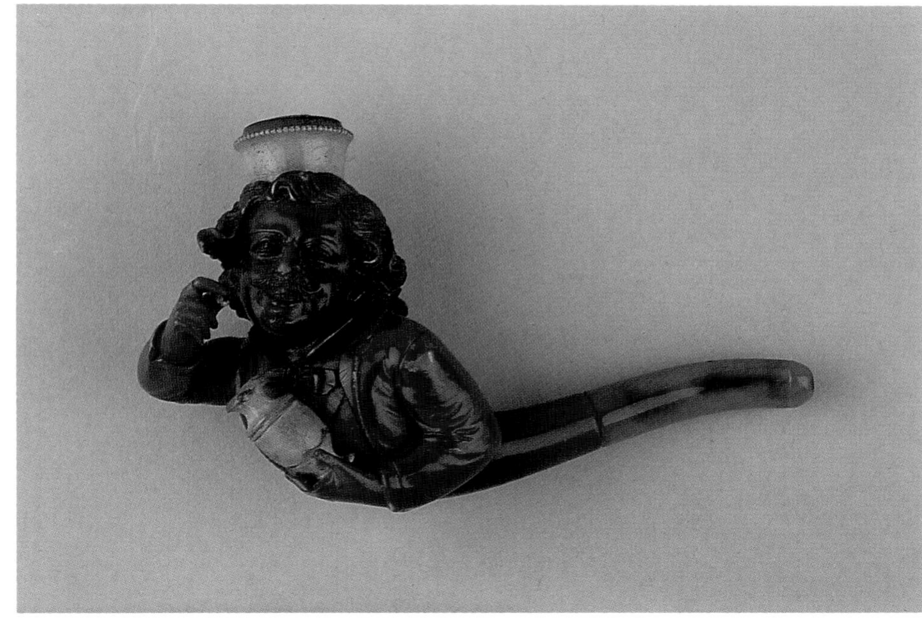

Pipe, pre-colored bust of gentleman dipping snuff from jar, 5.5" l., 3.5" h., C. W. Möller, Berlin, case embossed *"Nur Für Kenner"* (Only for Connoisseurs), ca. 1900. *Courtesy of the Author's Collection.*

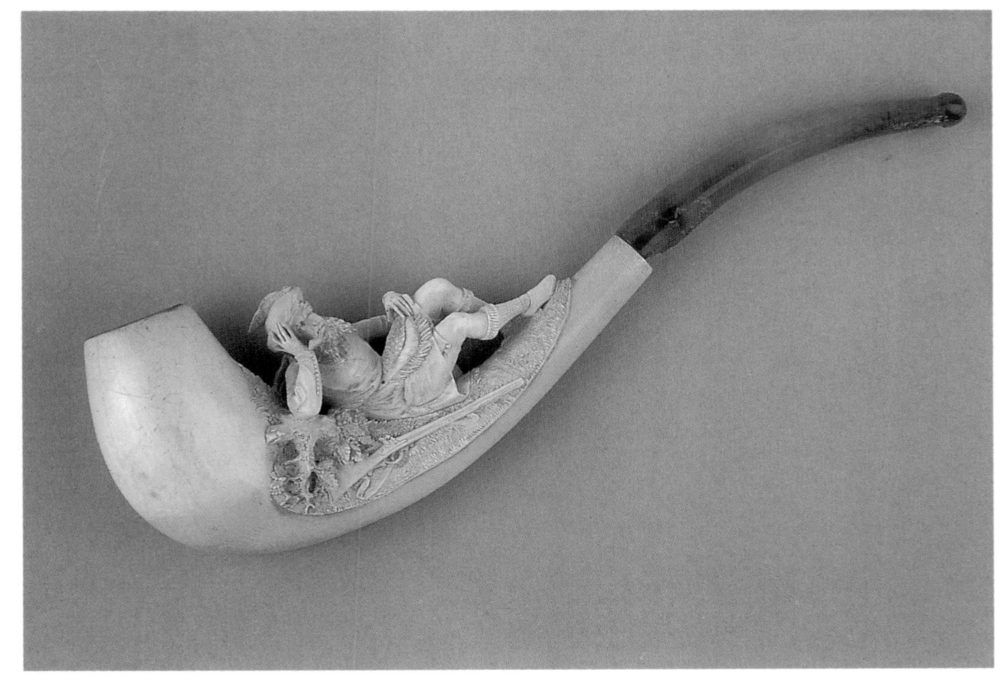

Pipe, kilted Scotsman as hunter at rest smoking pipe, rifle at side, 9" l., 3" h., Heinrich Schilling, Wien, ca. 1900. *Courtesy of the Author's Collection.*

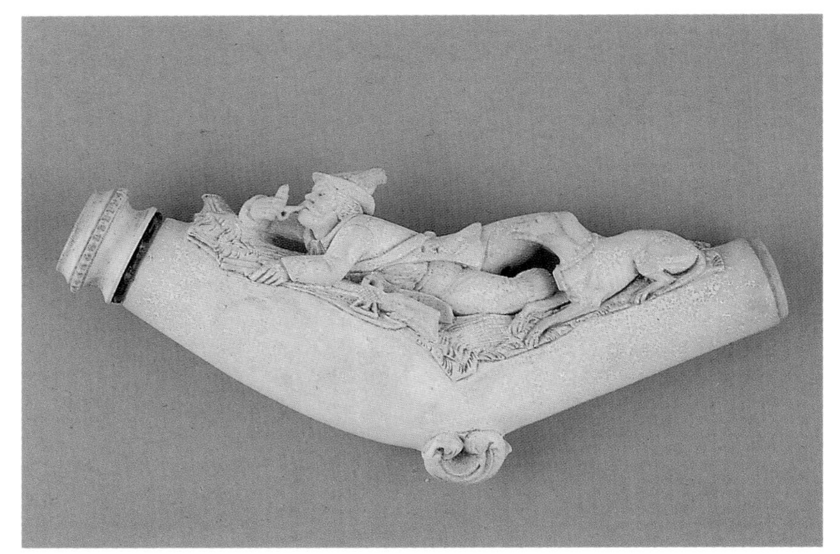

Cheroot holder bowl, hunter at rest, smoking pipe, 5.5" l., 1.5" h., J. C. Zangenberg, Osnabruck, ca. 1875. *Courtesy of the Author's Collection.*

Pipe, fisherman in stocking cap with his day's catch, smoking pipe, 11" l., 6.75" h., prob. French or German, ca. 1875. *Courtesy of the GC Collection.*

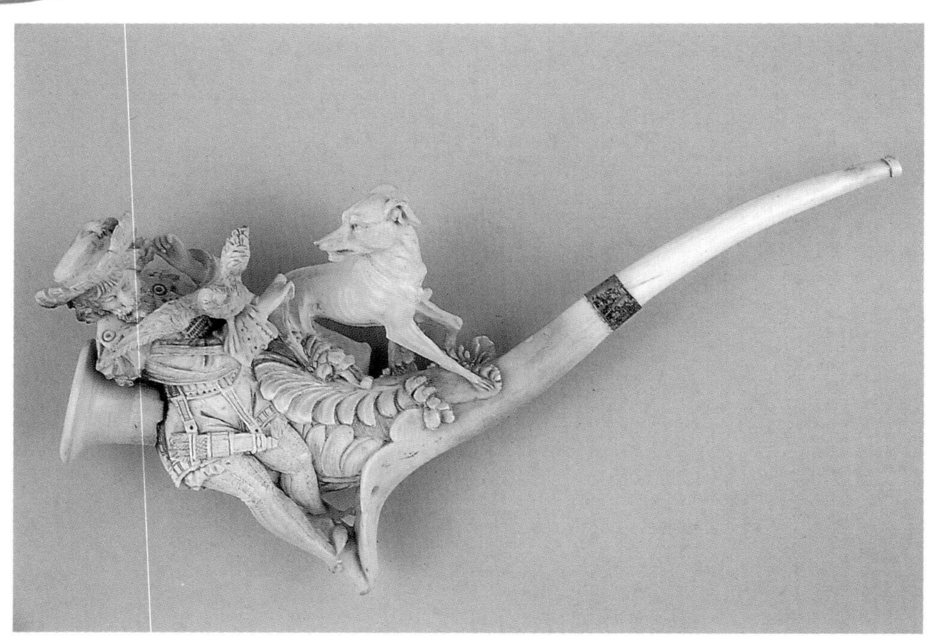

Cheroot holder, medieval falconer, accompanied by hound, 10.5" l., 5" h., prob. French, ca. 1875. *Courtesy of the SP Collection.*

Cheroot holder, hunter with game, a chamois-buck, and three hounds, 8" l., 4" h., Austrian-Hungarian, ca. 1875. *Courtesy of the GC Collection.*

Pipe, eagle's talon and egg-shaped bowl, silver and amber windcap, chased silver shank mount, 8" l., 4" h., L. Baumgartner, Trieste, ca. 1875. *Courtesy of the RHW Collection.*

Pipe, eagle's talon and egg-shaped bowl, alternating claws and figurative heads of fox, lion, and wolf circumscribe bowl, silver and amber windcap, chased silver shank mount, 9" l., 4" h., L. Baumgartner, Trieste, ca. 1875. *Courtesy of the Author's Collection.*

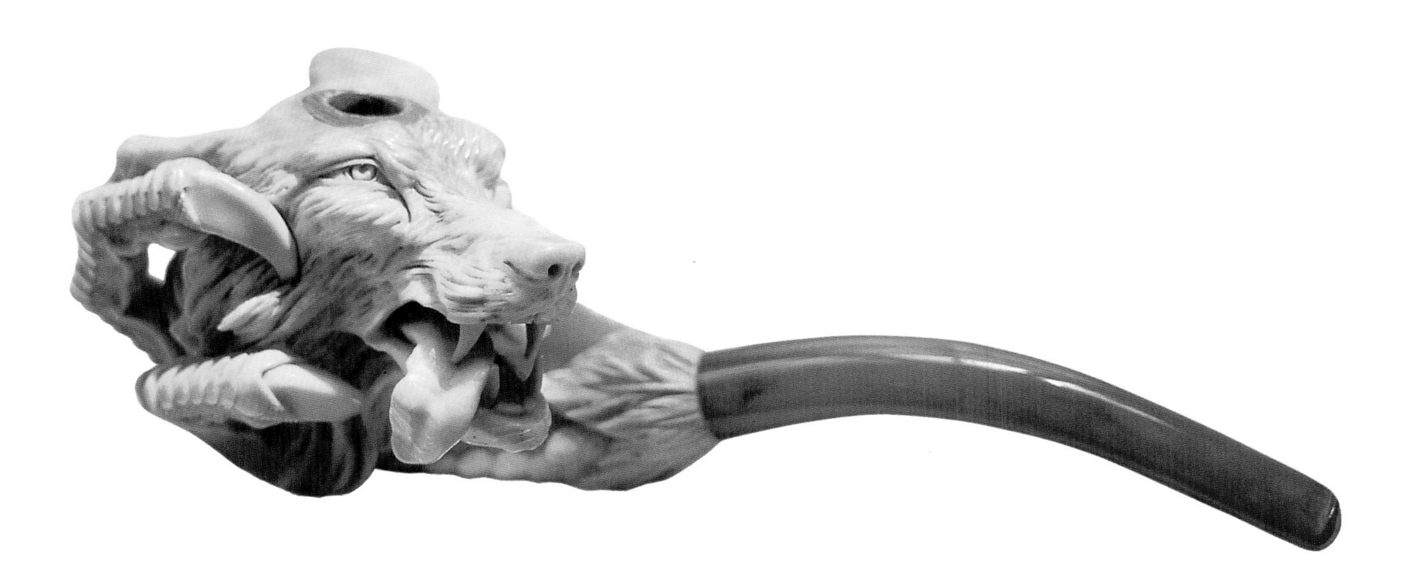

Pipe, figurative head of wild dog in the clutch of eagle's talon, 10.5" l., 3.5" h., prob. French, ca. 1885. *Courtesy of the MG Collection.*

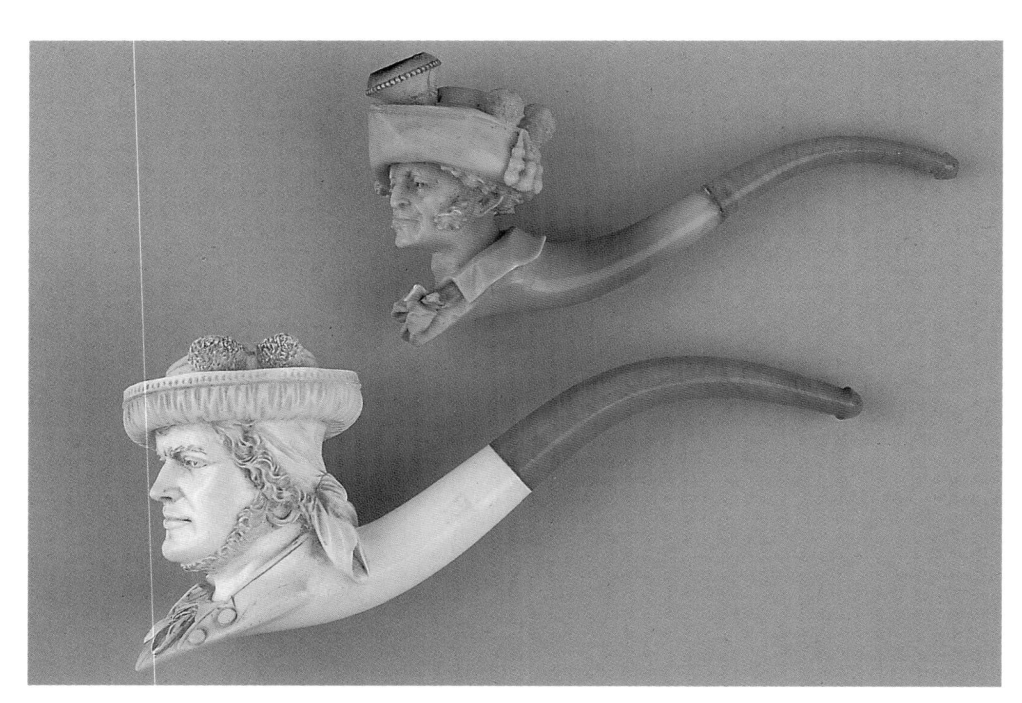

Cheroot and pipe, figurative head of picador with mutton chops, 5.5" l., 3" h., prob. French, ca. 1880 (top); figurative head of toreador or matador, 7.5" l., 3" h., WDC, American, ca. 1890 (bottom). *Courtesy of the Author's Collection.*

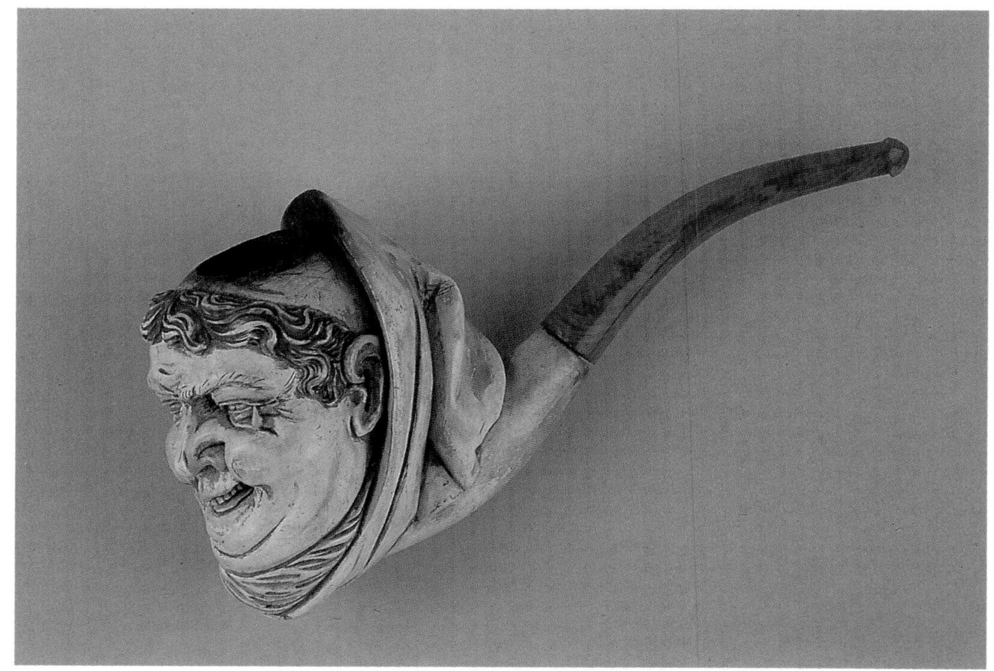

Pipe, figurative head of laughing monk in cowl, 7" l., 3" h., L.L. Stoddard, New Haven, Connecticut, ca. 1885. *Courtesy of the RC Collection.*

Pipe bowl, Pasha in traditional Near-East garb at tea time, seated on hassock, crossed dirks in waistband, silver accents on jacket, turban, and base, 8" l., 5.5" h., prob. Austrian Empire, ca. 1860. *Courtesy of the GC Collection.*

Pipe, masquerader or buffoon, after Pierrot, the French pantomime, 11" l., 5.5" h., J. Sommer, Paris, ca. 1900. *Courtesy of the MG Collection.*

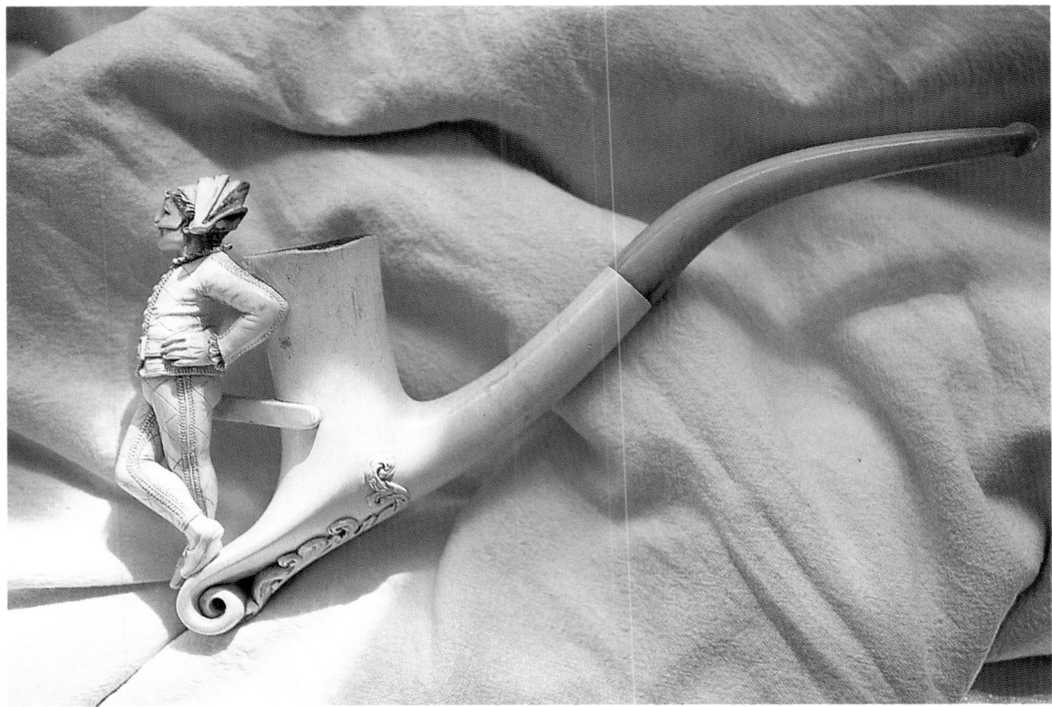

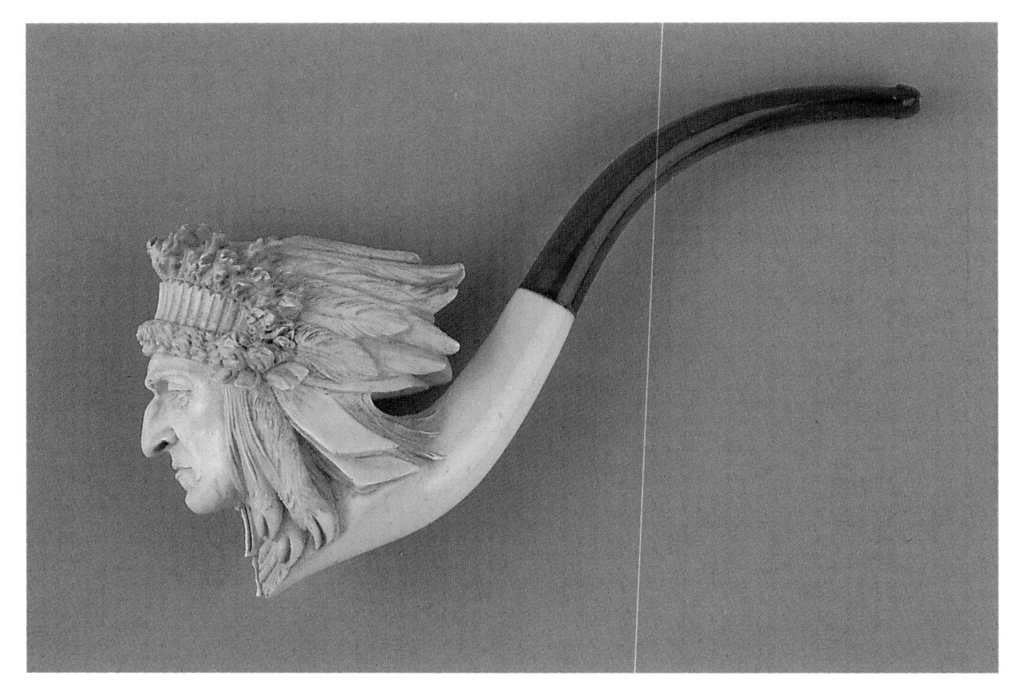

Pipe, figurative head of American Indian in headdress, 8" l., 3.75" h., prob. American, ca. 1890. *Courtesy of the AZ Collection.*

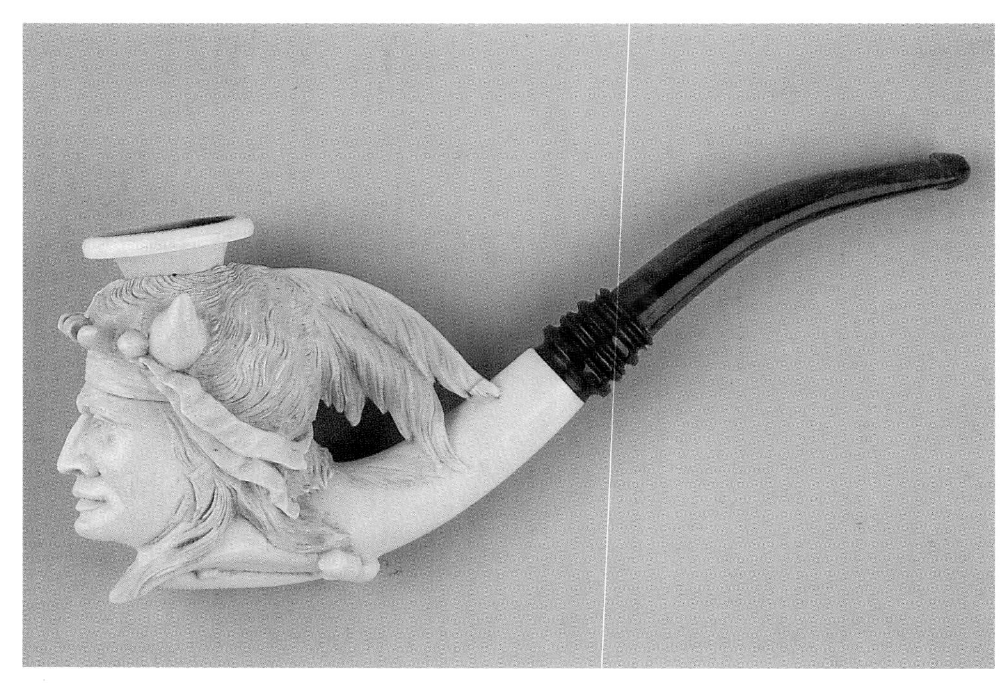

Pipe, figurative head of American Indian in headdress, 9" l., 4" h., underside of bowl displays raised decor of crossed tomahawk and peace pipe, prob. American, ca. 1890. *Courtesy of the RHW Collection.*

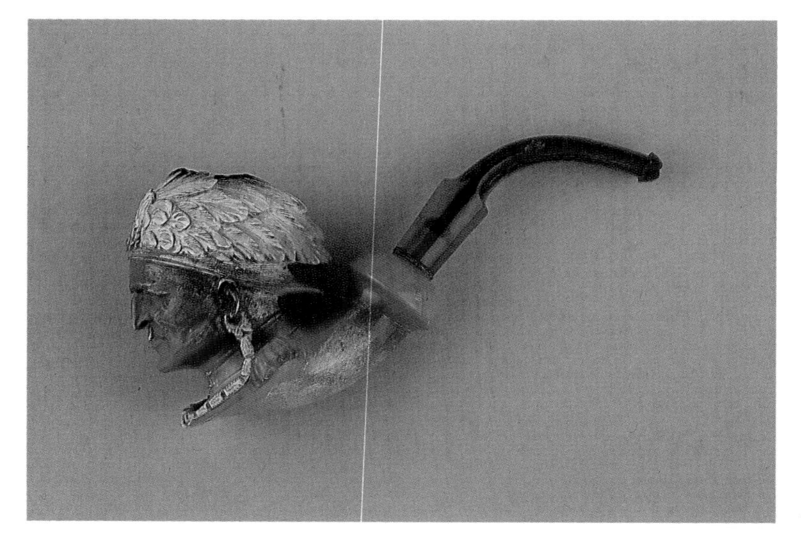

Pipe, figurative head of American Indian in headdress, 6" l., 3" h., prob. American, ca. 1885. *Courtesy of the Author's Collection.*

Pipe, figurative head of American Indian in headdress, bi-color accents, underside of bowl displays incised crossed arrows, prob. American, ca. 1890. *Courtesy of the Author's Collection.*

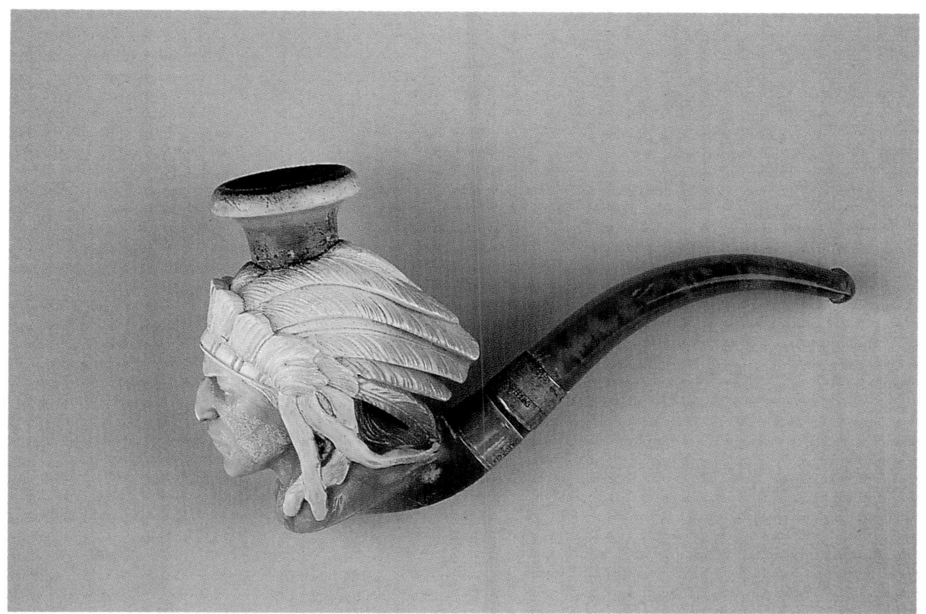

Pipe, figurative head of American Indian in headdress, 8" l., 4" h., prob. American, ca. 1890. *Courtesy of the FB Collection.*

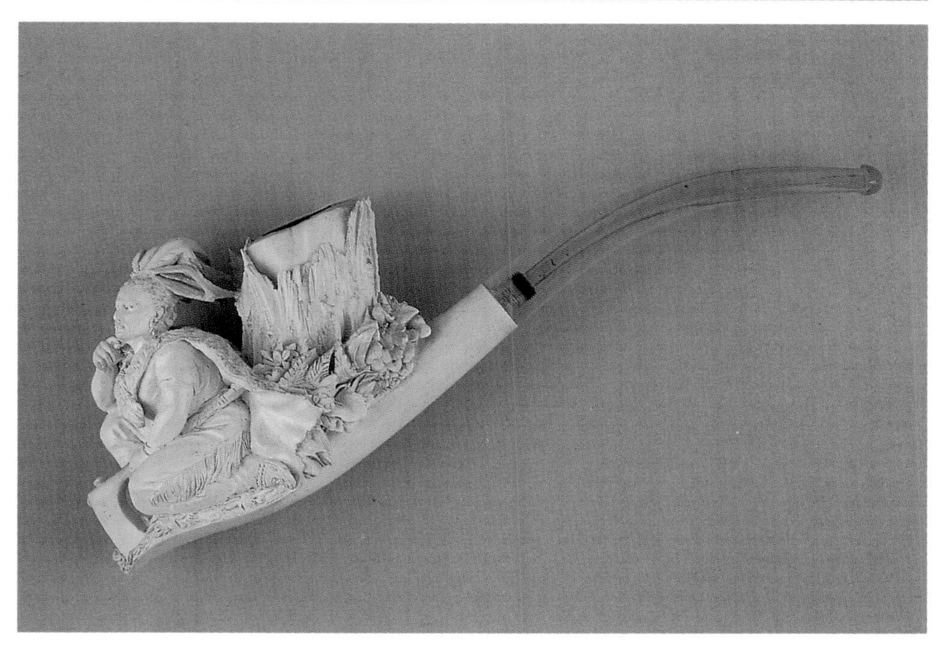

Pipe, American Indian in headdress on lookout by tree stump, tomahawk resting on left knee, 8.5" l., 4" h., prob. American, ca. 1880. *Courtesy of the SP Collection.*

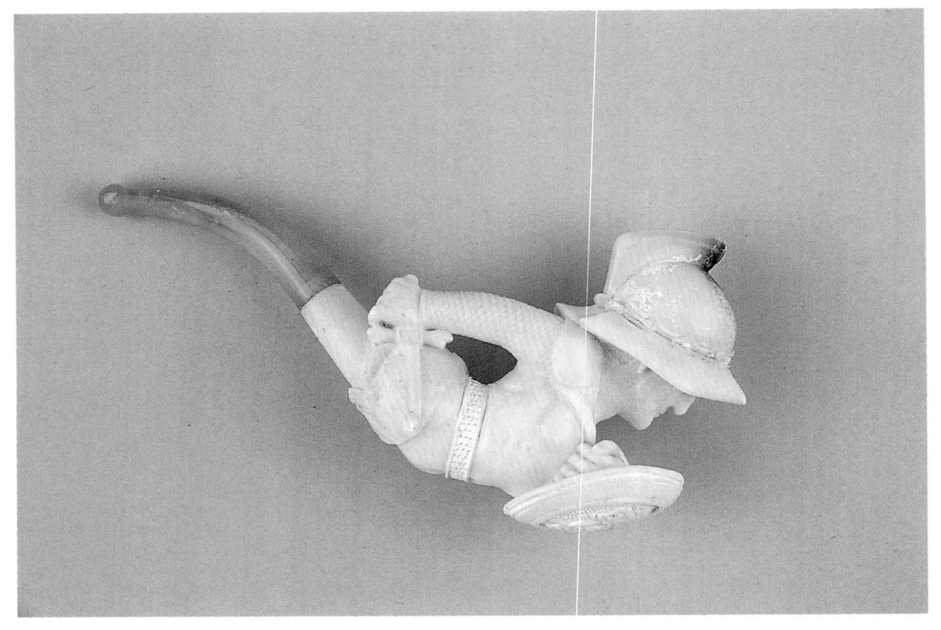

Pipe, unusual construct of gladiator in helmet, holding shield and sword, bi-color accents, 7.5" l., 3.5" h., prob. Italian, ca. 1900. *Courtesy of the SP Collection.*

Cheroot holders, cavalier blowing trumpet, 5.5" l., 2.5" h., prob. Austrian-Hungarian, ca. 1875 (top); bugler, perhaps sounding the commencement of the hunt, 5.5" l., 2.5" h., prob. German, ca. 1885 (bottom). *Courtesy of the Author's Collection.*

Cheroot holder, gaucho lassoing mare, 7" l., 2.5" h., prob. German, ca. 1885. *Courtesy of the AZ Collection.*

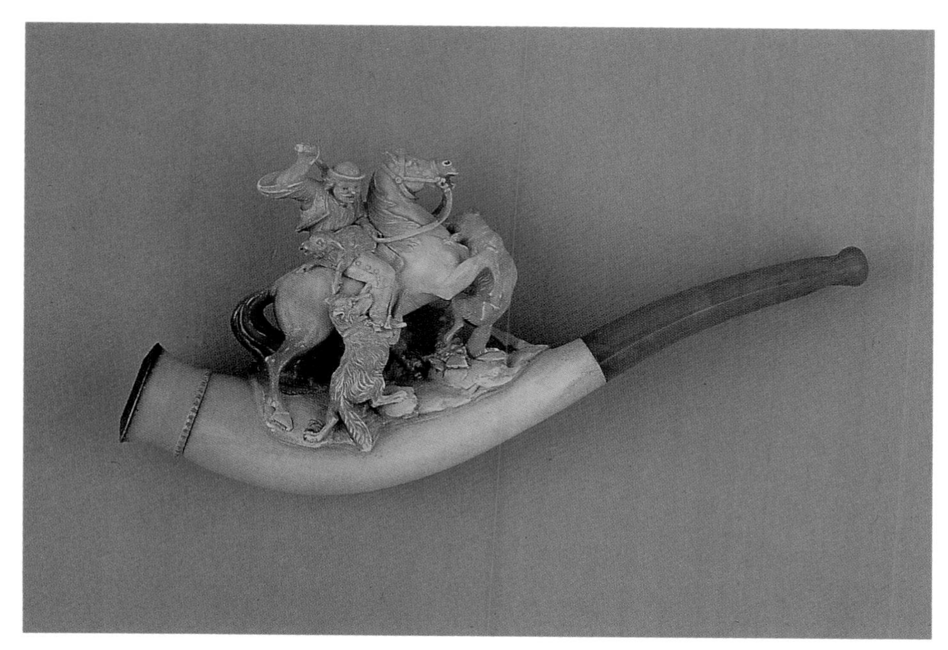

Cheroot holder, man on horse being attacked by cougar and wild dog, 6.5" l., 3" h., prob. German, ca. 1885. *Courtesy of the Author's Collection.*

Cheroot holder, three of Man's assorted best friends, 7.75" l., 3.25" h., prob. English or American, ca. 1890. *Courtesy of the AZ Collection.*

Cheroot holder, Sebastopol-style, group of four wild horses, 4" l., 2.75" h., prob. German, ca. 1870. *Courtesy of the GC Collection.*

Pipe, figurative head of horse with hame, 7.5" l., 3.5" h., Au Pacha, Paris, ca. 1880. *Courtesy of the Author's Collection.*

Cheroot holder, figurative head of horse with crown-like coloring bowl, 10.5" l., 5" h., prob. French, ca. 1885. *Courtesy of the Author's Collection.*

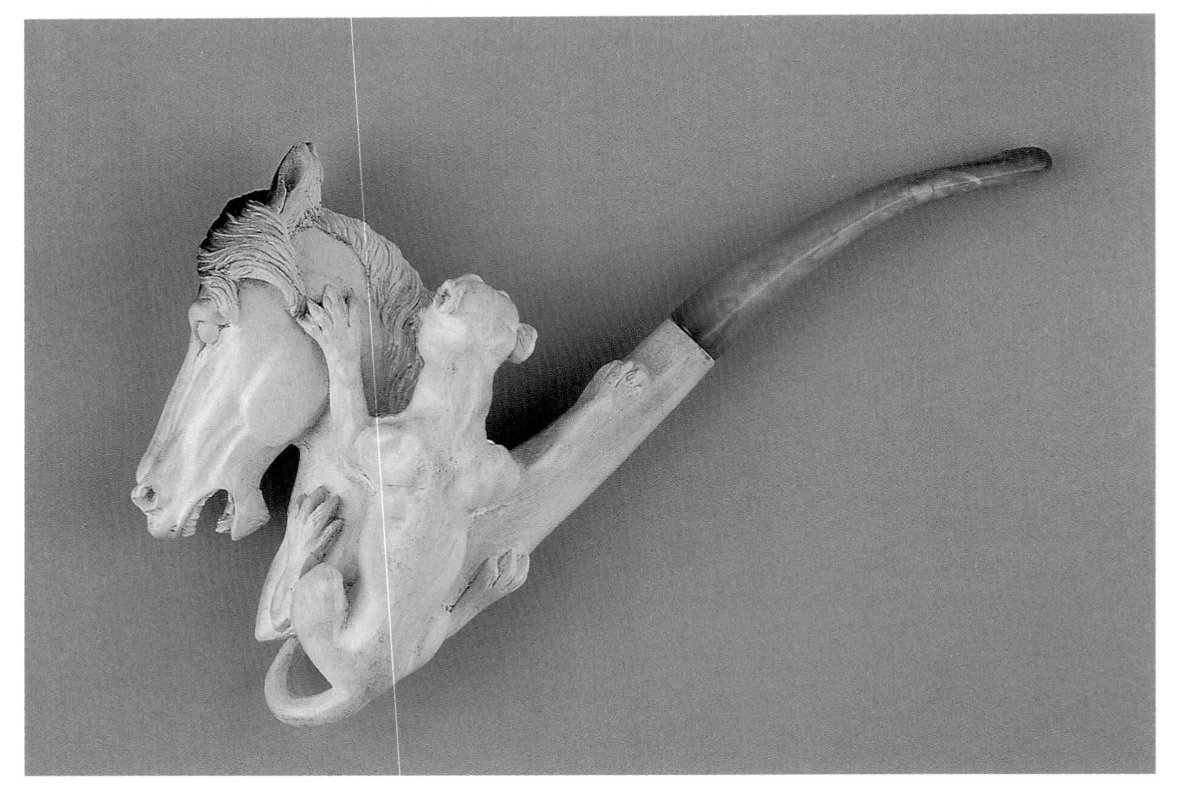

Pipe, figurative head of horse being attacked by panther, 7.75" l., 5" h., prob. Austrian, ca. 1880. *Courtesy of the SP Collection.*

Pipe, figurative head of horse being attacked by mountain lion, 17" l., 5" h., prob. Austrian-Hungarian, ca. 1890. *Courtesy of the GC Collection.*

Pipe, horse courant, tail mounted in silver windcap, ornately chased silver shank mount, 8.75" l., 4" h., prob. German, ca. 1890. *Courtesy of the GC Collection.*

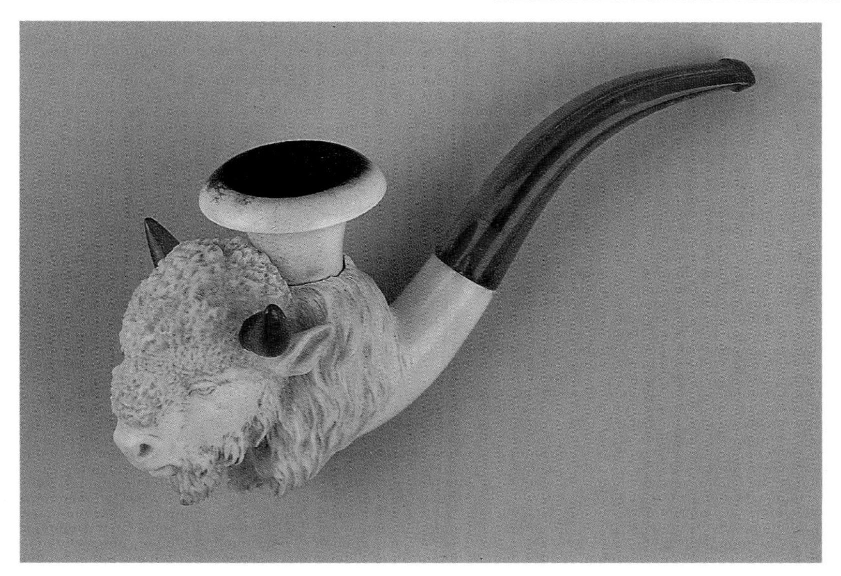

Pipe, figurative head of bison with amber horns, 7" l., 3" h., G. A. F., American, made for the Pan-American Exposition, 1901. *Courtesy of the Author's Collection.*

Pipe, figurative head of bull with amber horns, 8" l., 4" h., prob. American, ca. 1885. *Courtesy of the Author's Collection.*

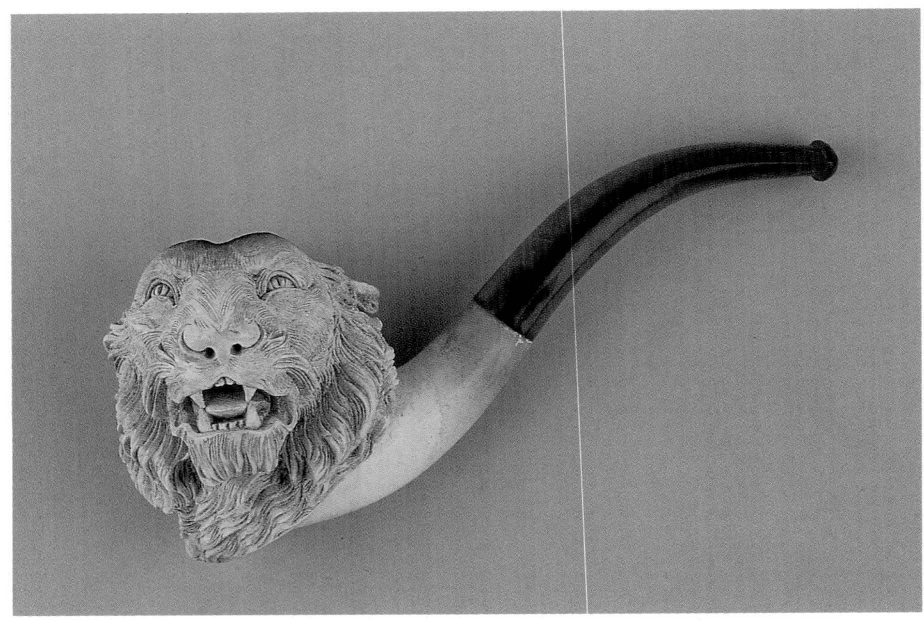

Pipe, figurative head of cougar, 6.5" l., 3" h., R. & W. Jenkinson Company, Pittsburgh, ca. 1885. *Courtesy of the RC Collection.*

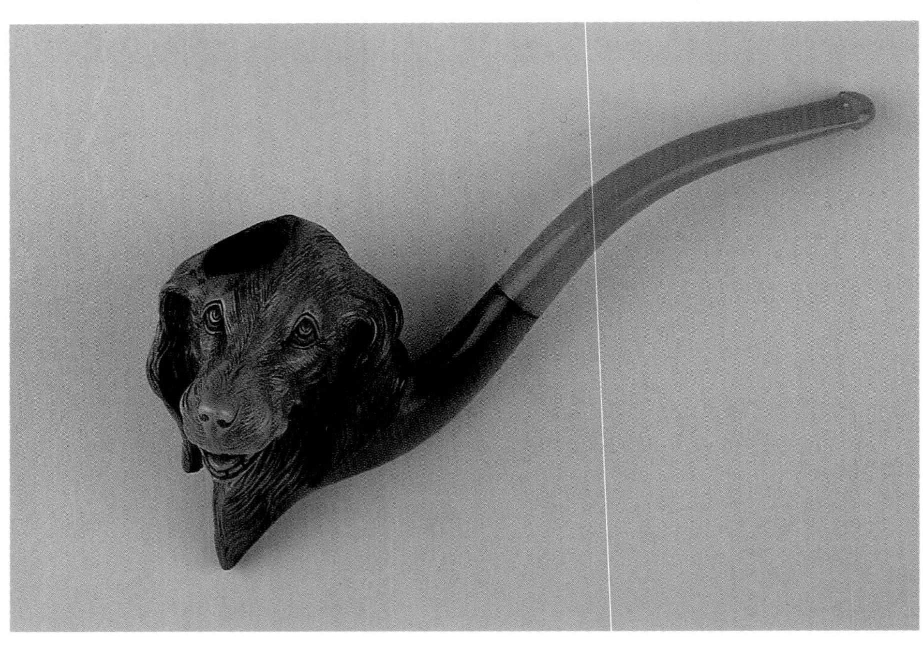

Pipe, pre-colored figurative head of setter, 7" l., 3" h., prob. American, ca. 1900. *Courtesy of the Author's Collection.*

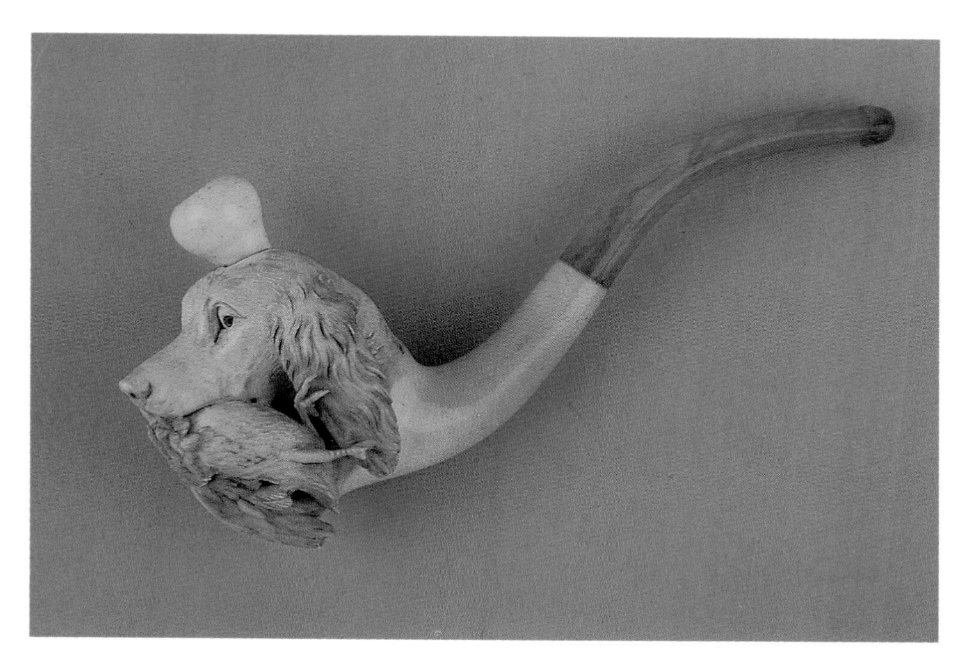

Cheroot holder, figurative head of hound with pheasant in mouth, 6" l., 3" h., prob. American, ca. 1900. *Courtesy of the Author's Collection.*

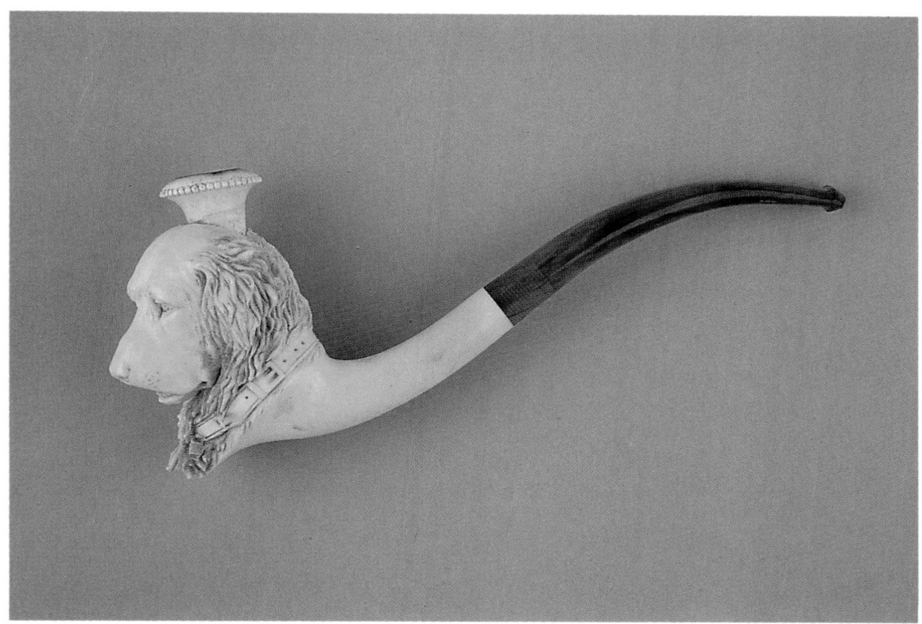

Pipe, figurative head of collared retriever, 7" l., 3" h., prob. American, ca. 1900. *Courtesy of the Author's Collection.*

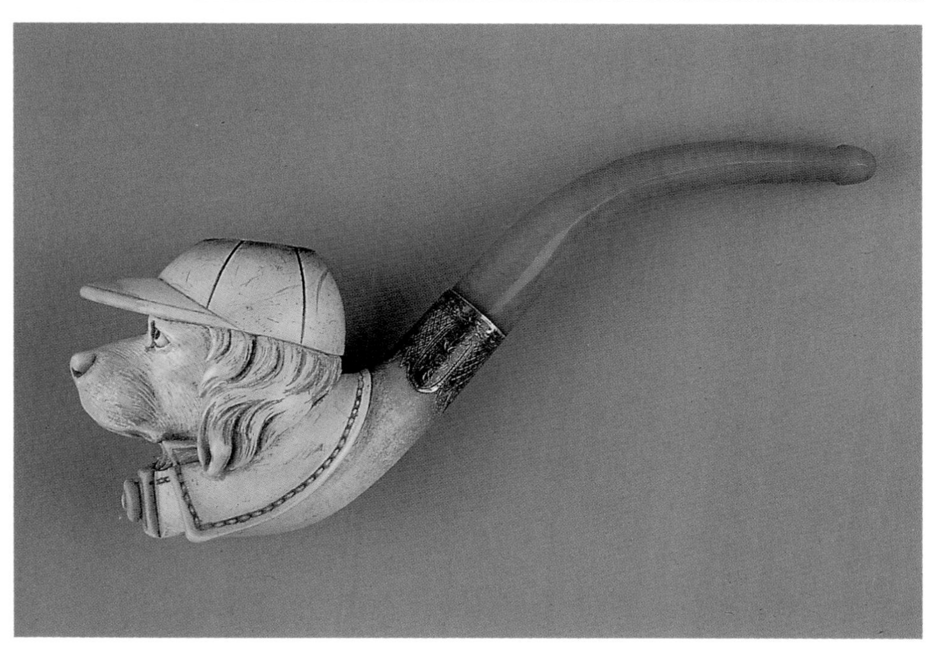

Pipe, figurative head of dog with cap, 7.5" l., 2.25" h., Au Pacha, Paris, ca. 1900. *Courtesy of the SP Collection.*

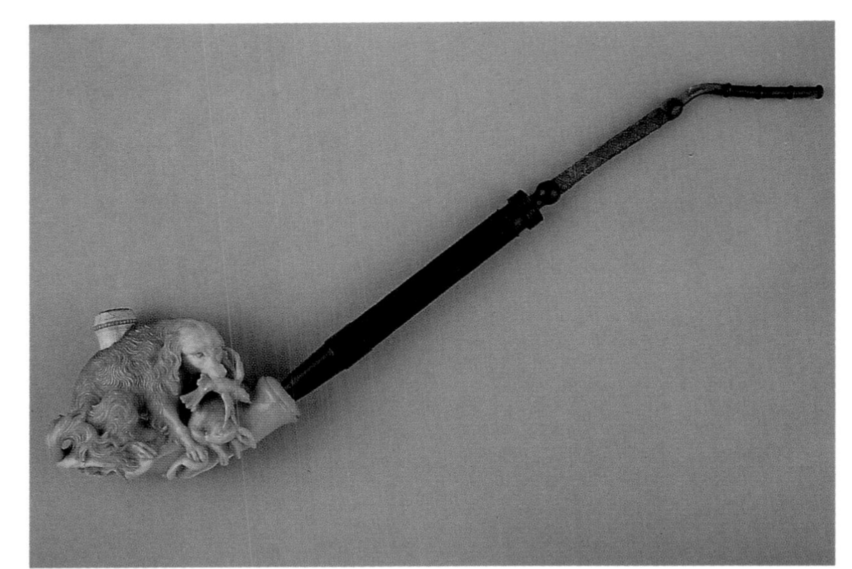

Pipe bowl, figurative head of hound with rabbit and bird quarry, 5.5" l., 4.5" h., accompanied by turned horn and flexible hose stem and horn mouthpiece. *Courtesy of the Author's Collection.*

Pipe bowl, Roman legionnaire on mount holding unit standard, 6.5" l., 5.5" h., prob. Italian, ca. 1870. *Courtesy of the FB Collection.*

Pipe bowl, lion sejant, 5.75" l., 7" h., prob. Italian, ca. 1870. *Courtesy of the SP Collection.*

Pipe, dog sejant, leash in mouth, 7" l., 5" h., prob. Italian or Austrian-Hungarian, ca. 1900. *Courtesy of a WC Collector.*

Pipe, dog sejant, right paw raised, 7.5" l., 4.5" h., prob. Italian or Austrian-Hungarian, ca. 1900. *Courtesy of the BB Collection.*

Pipe, dog sejant, right paw raised (notice ear variant from previous slide), 7.25" l., 4" h., McCoy Brothers, Vienna, ca. 1900. *Courtesy of a Midwest Collector.*

Pipe, cat sejant, inset glass eyes, 7" l., 5" h., G. Lightenstern, Milano, ca. 1900. *Courtesy of the DES Collection.*

Pipe, elephant and giraffe, 9" l., 3.75" h., prob. English, case marked "Vienna Exhibition 1873" (if the attribution is correct, this is an anomaly in the mutation chronology of the meerschaum). *Courtesy of the JTB Collection.*

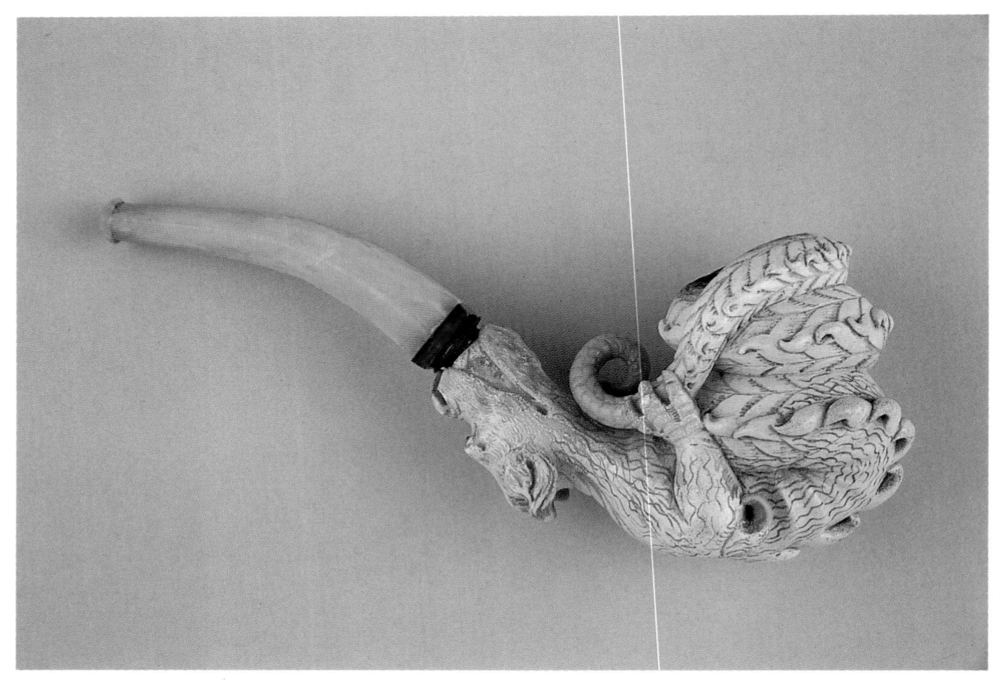

Pipe, figurative of inverted stylized dragon with spiraled tail, 7.25" l., 3.25" h., prob. French, ca. 1875. *Courtesy of the SP Collection.*

Chapter XII. Myth and Fable

Both myth and fable are fiction. The imaginary, the fictitious, and the legendary comprise the universe of the extraordinary and the supernatural, while fable includes allegory, parable, and invention. The two words may seem interchangeable, but Poseidon, the ancient god of the sea, was a hero in Greek mythology, whereas the tales of fabulists, such as the tortoise and the hare, are of a slightly different genre. Typically, fables taught a moral, more often than not, with animals or inanimate objects. Because of its craftiness, beauty, and solitary nature, the fox has figured prominently in fable whenever deceit, pride, or individuality was essential to the moral. The fox has played a significant role in every society that is familiar with this animal, beginning with the classic literature of Aesop written some 2,500 years before Joel Chandler Harris's American folklore stories of Uncle Remus and Bre'r Fox. For whatever reason, the fox, as is seen in this chapter, was also a prominent motif in meerschaum carvings.

However and why it occurred, the separate and dissimilar worlds of literature and meerschaum carving coalesced early in the 19th century, and the product of that merger was manifested in some very spectacular utensils for the smoker. Heroes and heroines who dramatized Greek and Roman mythology, and both the real and imagined animal role-players in fables can be found in picturesque scenes that circumscribe meerschaum pipes and rise above meerschaum holders. A unique array of mythical and fabulist subject matter is introduced in this chapter as proof that the opportunities for collecting are literally endless.

Cheroot holder, pre-colored, St. George slaying the dragon, 7.5" l., 4.5" h., Fr. Rosenstiel, Berlin, ca. 1880. *Courtesy of the Author's Collection.*

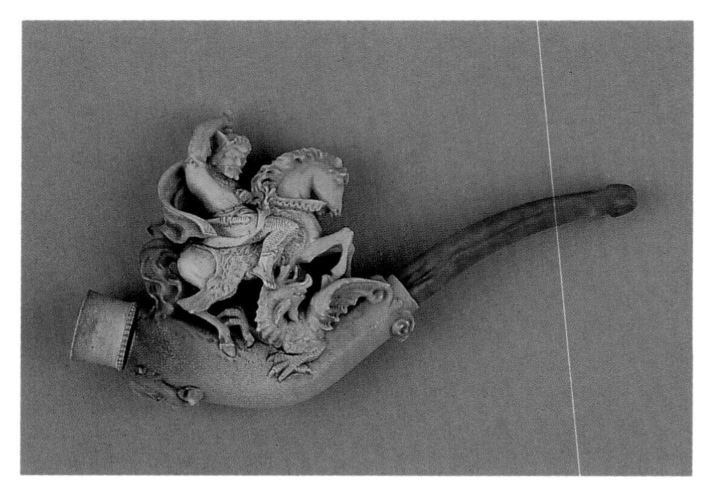

Cheroot holder, another version of St. George slaying the dragon, 5.5" l., 3.25" h., prob. German, ca. 1880. *Courtesy of the AZ Collection.*

Cheroot holder, a very young St. George and the fiery dragon, 6.5" l., 2.75" h., prob. German, ca. 1885. *Courtesy of the MG Collection.*

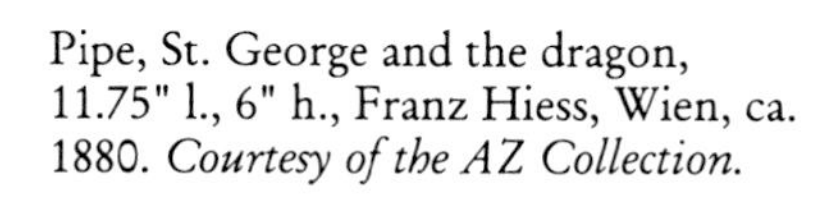

Pipe, St. George and the dragon, 11.75" l., 6" h., Franz Hiess, Wien, ca. 1880. *Courtesy of the AZ Collection.*

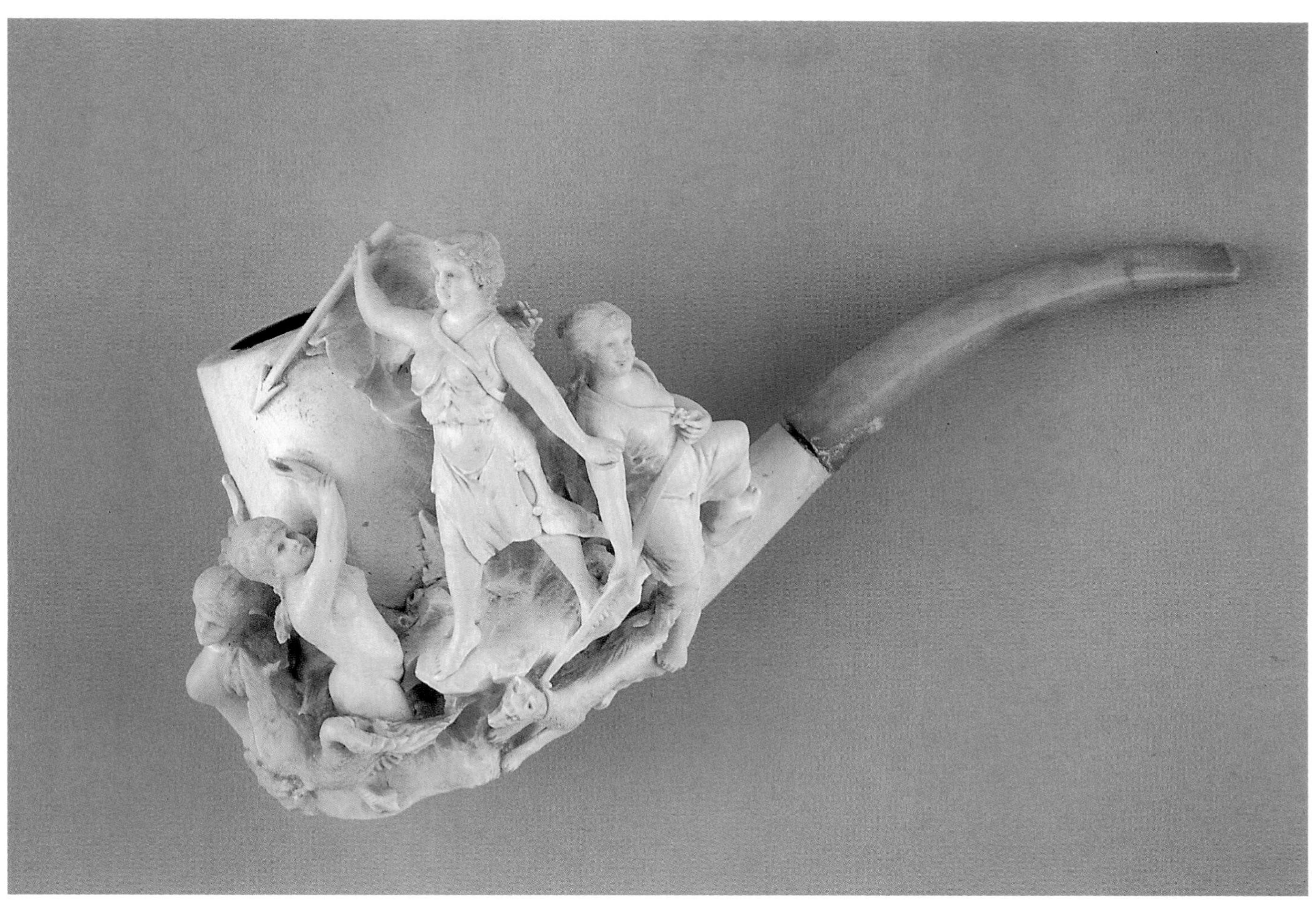

Pipe, in-the-round scene, the huntress Diana and five consorts in various poses, dog, deer and fowl in attendance, 9.5" l., 5" h., prob. Austrian-Hungarian, ca. 1880. *Courtesy of the FB Collection.*

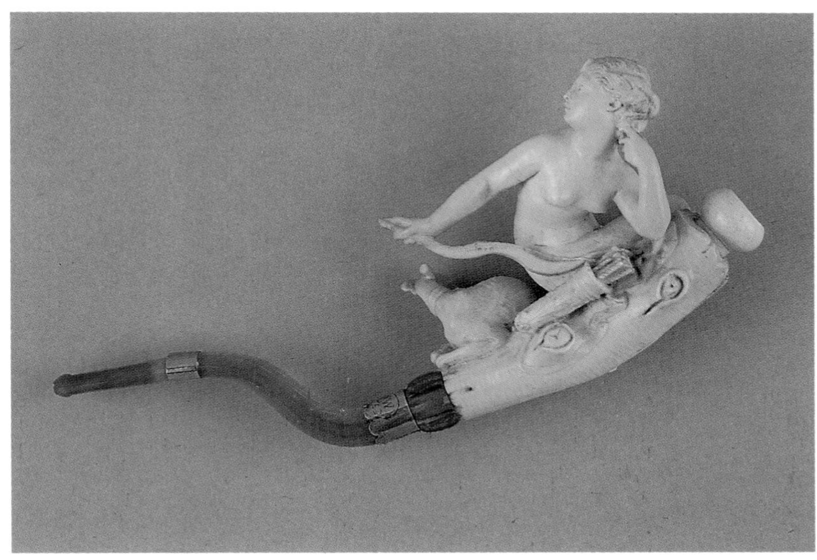

Right:
Cheroot holder, Diana with bow and quiver at rest, deer in attendance, 7.5" l., 4" h., L. Goetsch, Paris, ca. 1885. *Courtesy of the SP Collection.*

Cheroot holder, Leda and the Swan (Zeus), who parented Pollux and Helen, 6.5" l., 2.5" h., Au Pacha, Paris, ca. 1880. *Courtesy of the Author's Collection.*

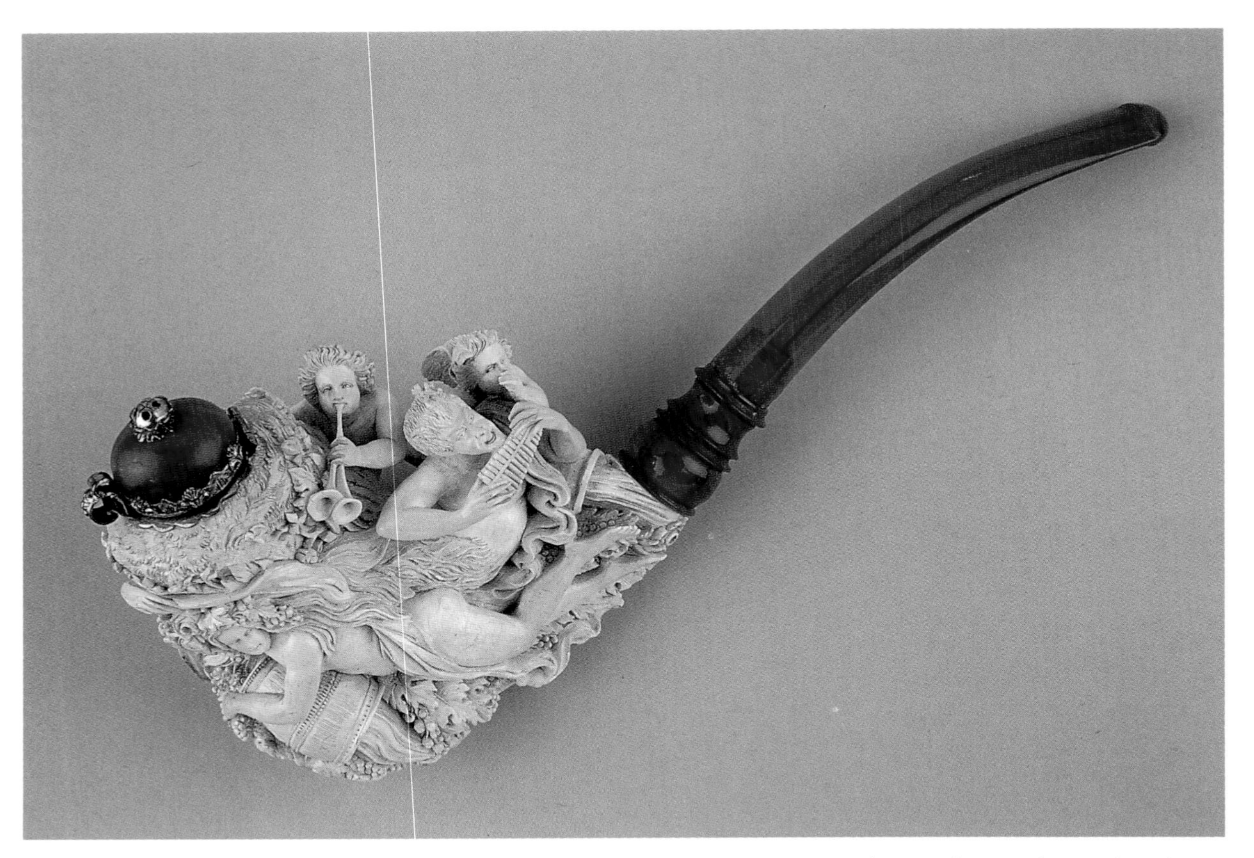

Pipe, in-the-round scene, the Greek god, Pan, god of goats, sheep and shepherds, and of music, playing the panpipe (reed), accompanied by other musicians. Pan was also known as the god of nature—meadows, forests, and beasts—and a womanizer, so he is customarily depicted in bucolic scenes with nudes and cherubim, as shown, 9.75" l., 4" h., Austrian-Hungarian, ca. 1875. *Courtesy of the FB Collection.*

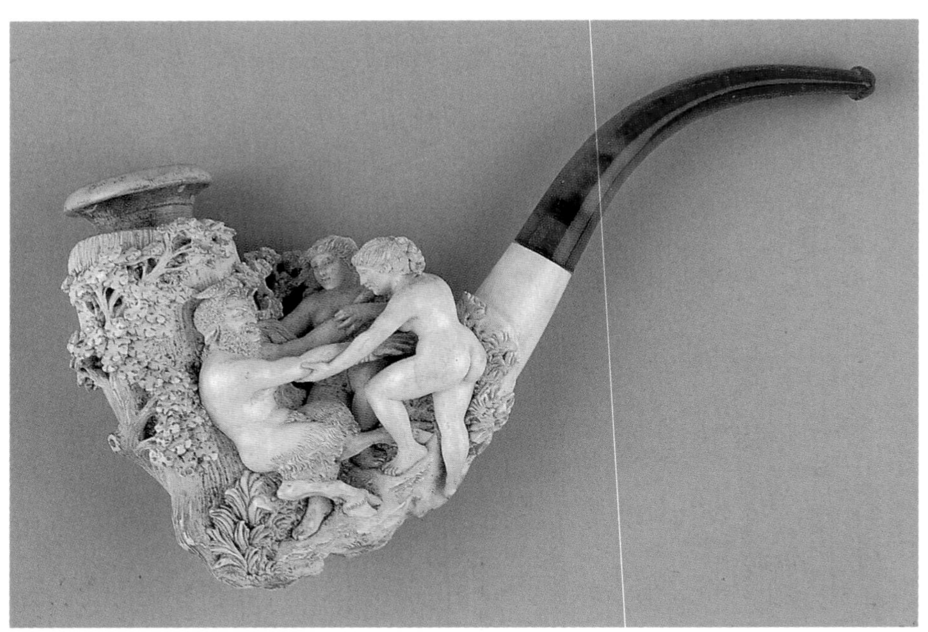

Left:
Pipe, another version of Pan frolicking in the wild with nudes, after a painting by Adolphe William Bouguereau, 8.5" l., 5" h., prob. German, ca. 1875. *Courtesy of the SP Collection.*

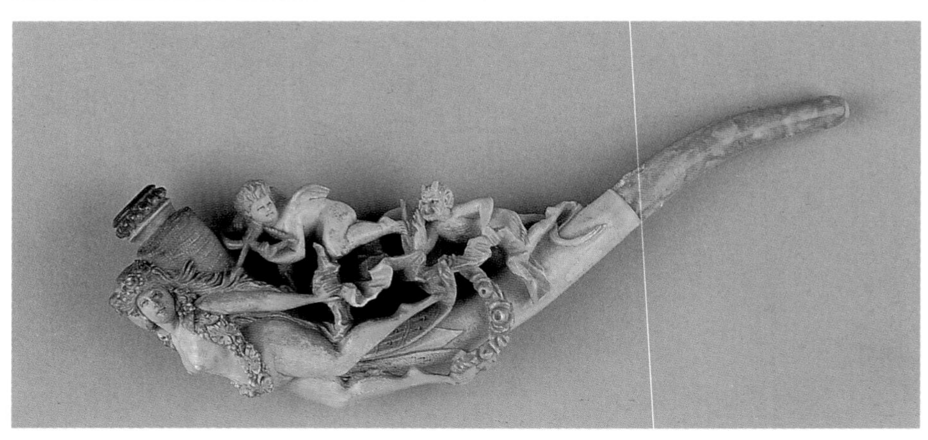

Cheroot holder, a more liberal rendition of Pan and friends, 8.5" l., 4" h., prob. German, ca. 1880. *Courtesy of the FB Collection.*

Pipe, Zeus (?) seducing nude on his horse-drawn chariot, 10" l., 4.5" h., case stamped *"Dem Fortschritte, 1873"* (exhibited at the *Weltausstellung, Wien*), prob. Austrian-Hungarian. *Courtesy of the Author's Collection.*

Below:
Poseidon had a fiery temper and many love affairs, especially with nymphs of springs and fountains. This pipe embodies his character, carved-in-the-round, sea horses and nereids, silver-gilt domed windcap inset with coral, meerschaum finial of Poseidon at rest in a grotto, chased silver-gilt shank mount inset with coral, 13" l., 8.75" h., prob. Austrian-Hungarian, ca. 1875. *Courtesy of the JTB Collection.*

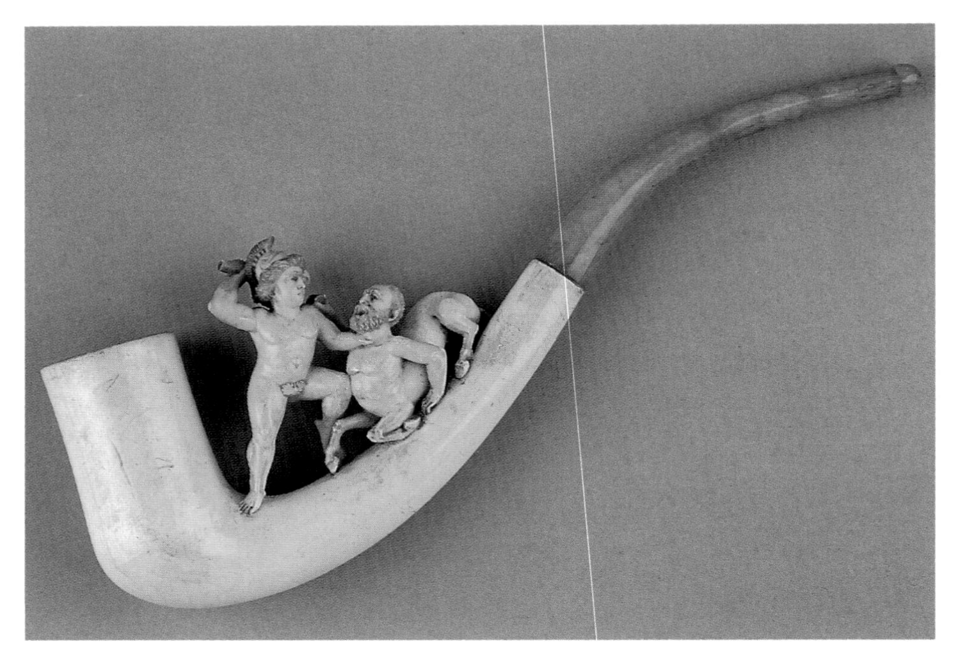

Theseus, the Attic hero, had many adventures, including battles against the Amazons and the Centaurs, the latter, battling a Sagittary, depicted on this pipe, 10" l., 4" h., J. Sommer Fabrique, Paris, ca. 1900. (Perhaps after the bronze by Antoine-Louis Barye, *Theseus Slaying Centaur*,1850.) *Courtesy of the SP Collection.*

As the mythology relates, Ganymede, a Trojan youth, was abducted and taken to Olympus where Zeus made him immortal. Rembrandt and Correggio paintings depicted the abductor as an eagle, whereas this pipe carver articulated the abductor as a vulture, 8" l., 4" h., prob. Austrian-Hungarian, ca. 1885. *Courtesy of the Author's Collection.*

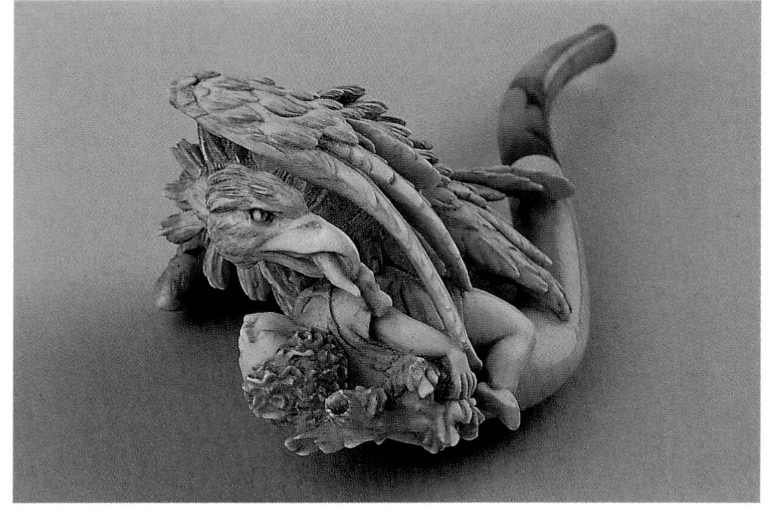

Below:
Cheroot holder, Athena, the Greek goddess of wisdom, fertility, the useful arts, and prudent warfare, as charioteer, 14" l., 4" h., Gebrüder Muschweck, Dresden, ca. 1880. *Courtesy of the SP Collection.*

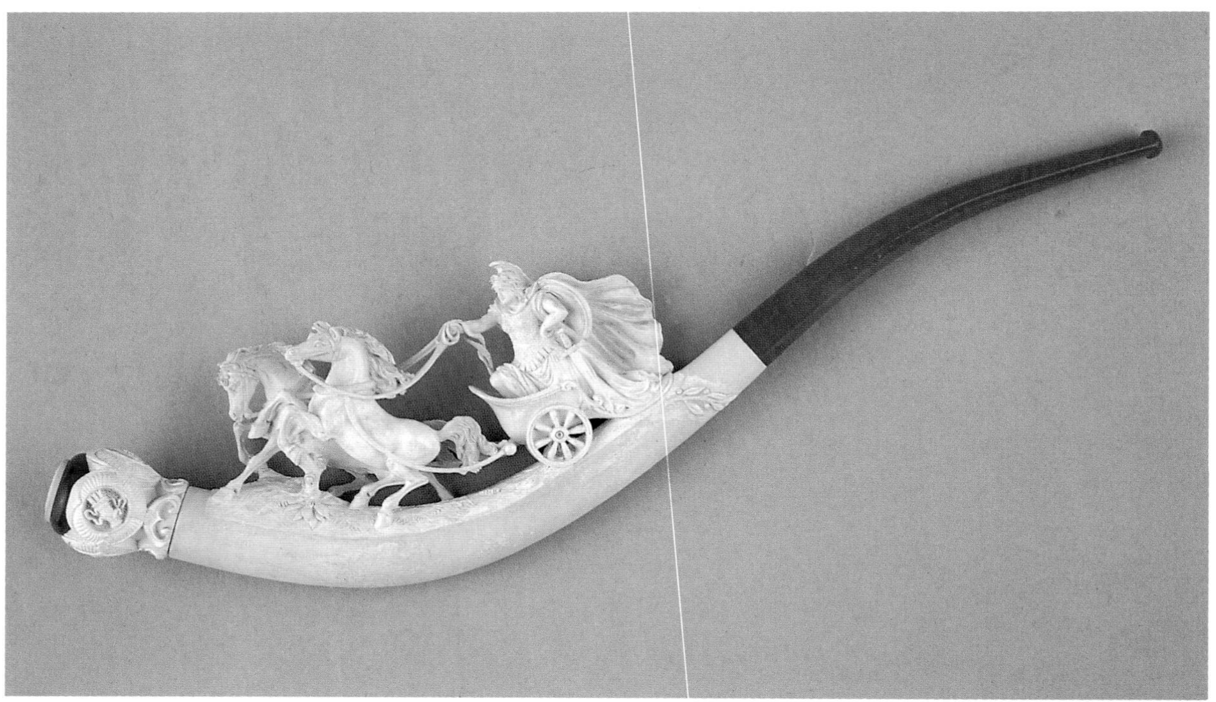

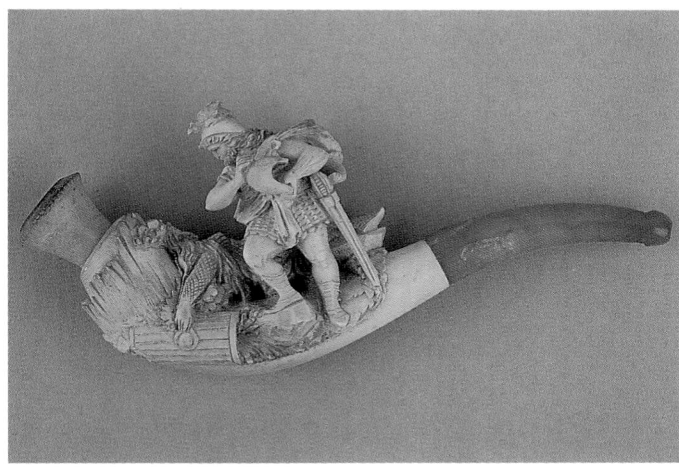

Cheroot holder, Brünnhilde, Valkyrie daughter of Wotan and Erda, and the mortal son of Wotan, Siegmund, from Richard Wagner's *Die Walküre, Der Ring des Nibelungen,* 8.75" l., 3.5" h., case stamped Gustav A. Fischer, Boston, ca. 1875. *Courtesy of the FB Collection.*

Cheroot holder, dramatic scene from Wagner's Die Walküre, 5.5" l., 2.5" h., Ostpreussische Bernstein Industrie, Berlin, ca. 1885. *Courtesy of the Author's Collection.*

Cheroot holder, Wagner's Wotan and Brünnhilde surmounted, 7.75" l., 2.75" h., Ecume Veritable, Paris, ca. 1880. *Courtesy of the SP Collection.*

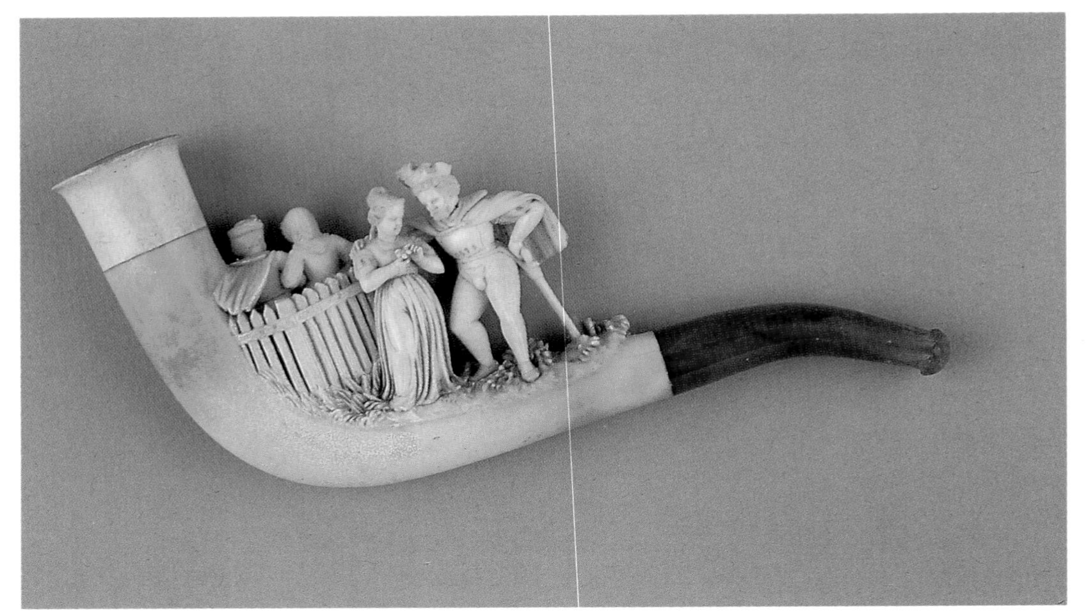

In the tragedy by Goethe (1831), and in the opera by Charles Gounod (1859), Johann Faust sells his soul to the devil for knowledge and power. This cheroot holder portrays Faust, Marguerite, Méphistophélès, and Marthe Schwerlein, Marguerite's neighbor, 7.5" l., 2.75" h., prob. German or French, ca. 1875. *Courtesy of the SP Collection.*

Cheroot holder, depiction of the beheading of Lady De Winter, Athos' ex-wife, from the Alexandre Dumas novel, *The Three Musketeers* (1844), 7.75" l., 2.75" h., prob. French, ca. 1880. *Courtesy of the SP Collection.*

Cheroot holder, Gambrinus, mythical Flemish king , reputed inventor of beer, and patron of all brewers, accompanied by the customary glass or mug, 5.5" l., 3" h., Emanuel Czapek, Prag, ca. 1875. *Courtesy of the Author's Collection.*

Cheroot holder, another rendition of Gambrinus, 5" l., 3" h., prob. Austrian-Hungarian, 1875. *Courtesy of the Author's Collection.*

Reynard the Fox of Waasland (Van den Vos Reynaerde ([Flanders]; Roman de Renart [Alsace-Lorraine]; and Reineke Fuchs [German]), hero of medieval beast epics, written about 1200, was also the subject of a Johann Wolfgang von Goethe epic poem. Illustration, William V. Kaulbach, *Reynard The Fox*, 1870, 4. This fairy tale about a not-so-cunning fox incentivized several meerschaum carvers. *Courtesy of the SP Collection.*

Cheroot holder, Reineke lazing against a tree trunk, his wife peeking out from the opposite end, 5" l., 2" h., prob. German, ca. 1880. *Courtesy of the SP Collection.*

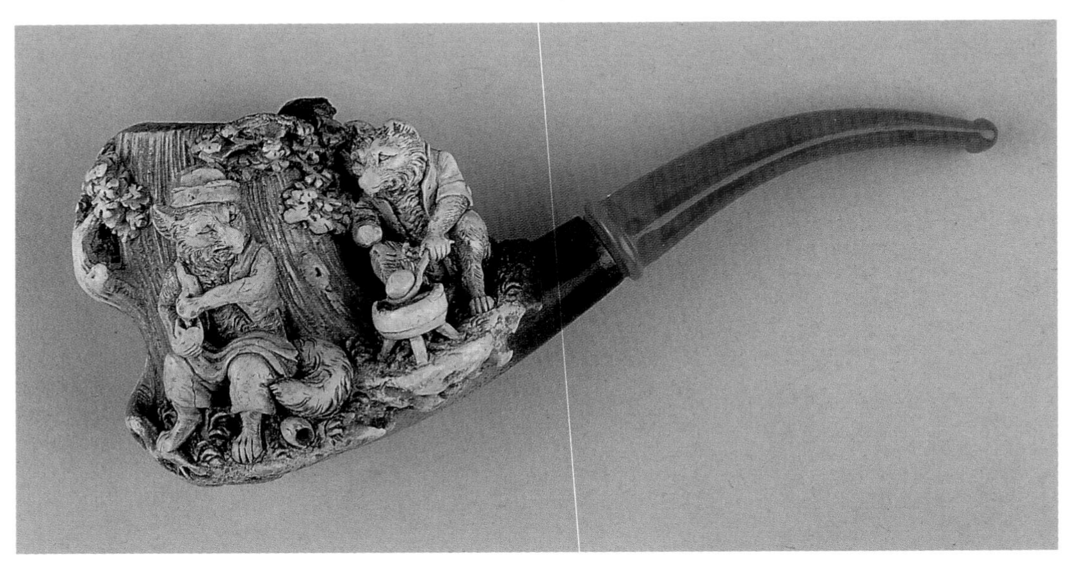

Pipe, Reineke and (prob.) Isengrin, the wolf, preparing a meal, 6.5" l., 2.75" h., prob. German, ca. 1875. *Courtesy of the SP Collection.*

Cheroot holder, a meeting of Reineke's uncle, Bruun, a bear, and others from the animal world, 5" l., 2.75" h., prob. German, ca. 1875. *Courtesy of the SP Collection.*

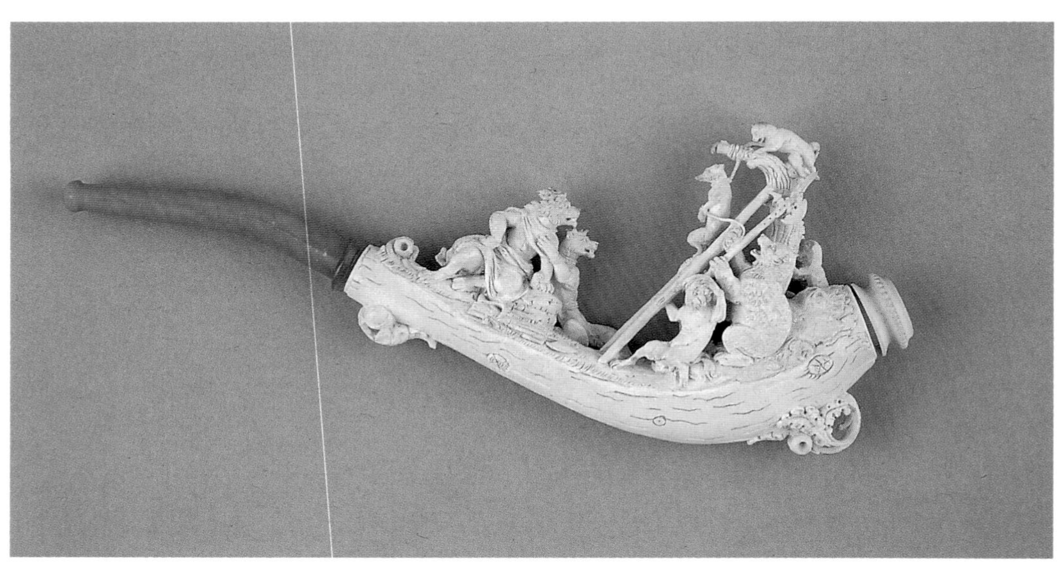

Cheroot holder, best of series, an exacting and diligently crafted replica of the final act, as Nobel, the lion king, oversees the demise of Reineke, while others of the animal kingdom observe, 7.75" l., 3" h., prob. German, ca. 1880. (This rendition appears on page 69 of Kaulbach's book.) *Courtesy of the SP Collection.*

Chapter XIII. Naiads, Putti, Cherubs, and Seraphim

Collectively, this chapter focuses on illustrations of the little people—infants, water nymphs and naiads, and the classic fine-art representations of celestial angels and winged cupids. Why devote a chapter to this genre? First, exclusive of nymphs and children, the remainder are symbolic of Eros, and second, as a group, I consider them to be excellent examples of the ideal diminutive in meerschaum. All delicate, intricate, and expressive, this genre is captivating and charming, and I hope that you react similarly to these images.

The concept of portraying real children in meerschaum must not have been considered an appropriate theme, because not many are in circulation today. Winged cupids accompanied by the standard accouterments of a bow and a quiver of arrows, however, must have been a more acceptable motif, because these are found with relative regularity.

Cheroot holder, nude with attendant putto standing on pedestal, 6" l., 3.5" h., G.W. Möller, Berlin, ca. 1890. *Courtesy of the Author's Collection.*

Cheroot holders, cupid playing flute, 5" l., 3" h., M. Czapek, Prag, ca. 1890 (top); cupid with bow and arrow, 6" l., 2.5" h., Ludwig Hartmann & Eidam, Wien, ca. 1890 (bottom). *Courtesy of the Author's Collection.*

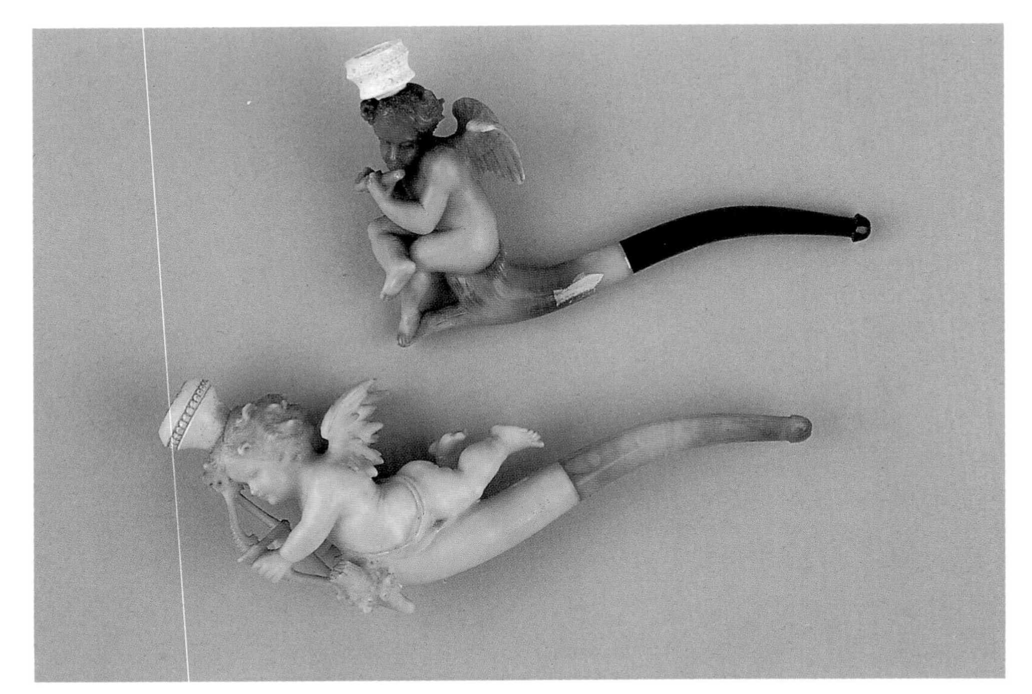

Cheroot holder, cupid drawing wagon bedecked with flowers, 4.5" l., 2.5" h., prob. German, ca. 1890. *Courtesy of the SP Collection.*

Cheroot holder, Mercury, the winged messenger, as young boy astride wheel, 4" l., 2" h., prob. French, ca. 1890. *Courtesy of the SP Collection.*

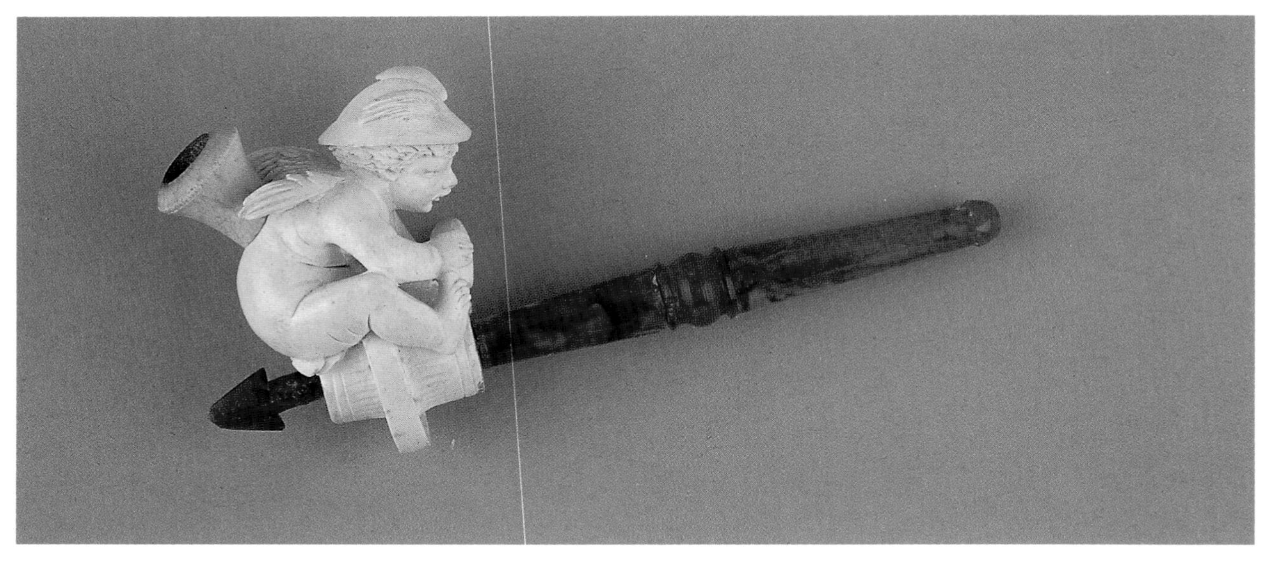

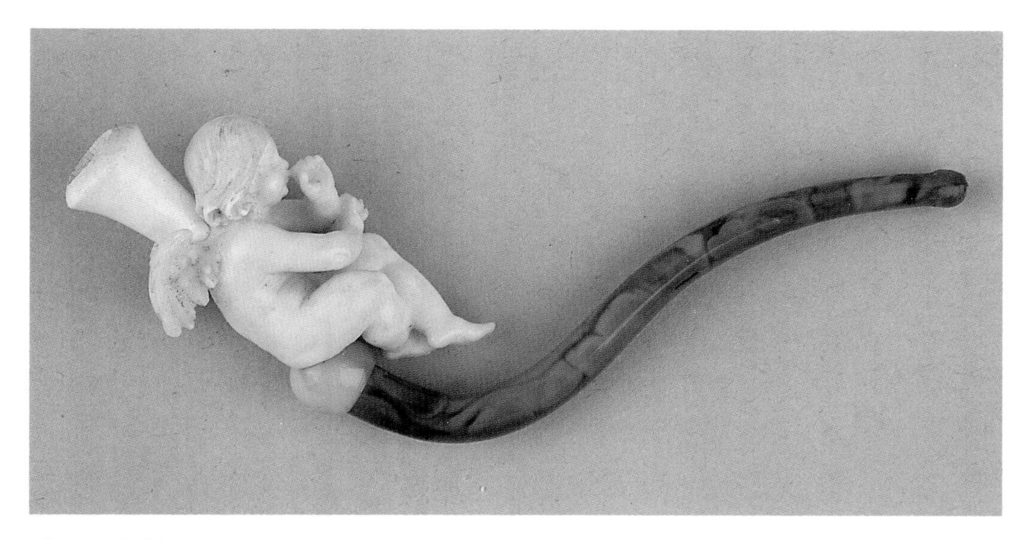

Cheroot holder, contemplative putto, 5" l., 2.5" h., prob. French, ca. 1890. *Courtesy of the Author's Collection.*

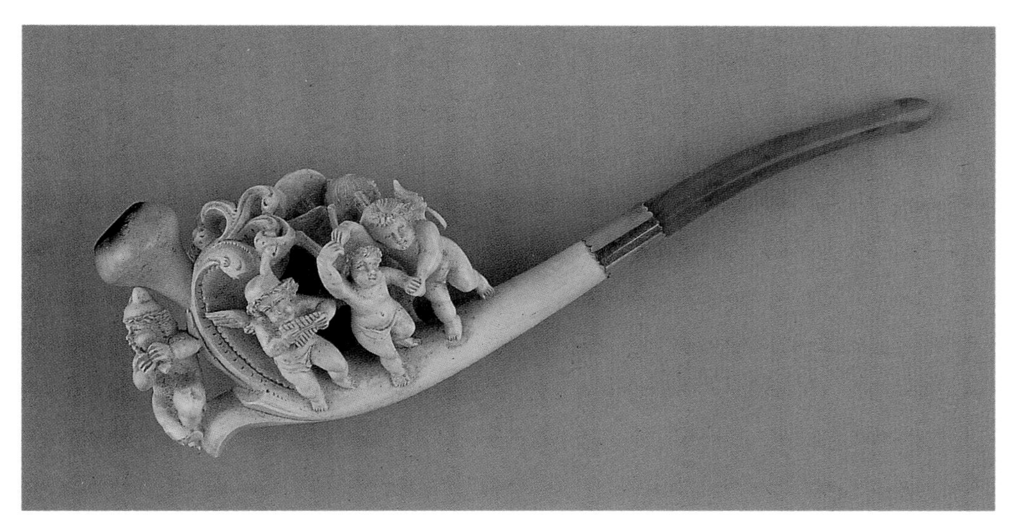

Cheroot holder, celestial band of merry musicians, 7" l., 2.5" h., Maurizio Fürst, Torino, ca. 1880. *Courtesy of the SP Collection.*

Cheroot holder, puttos playing heavenly instruments, 7" l., 3" h., William Astley, 109 Jermyn St., London, S. W., ca. 1880. *Courtesy of the FB Collection.*

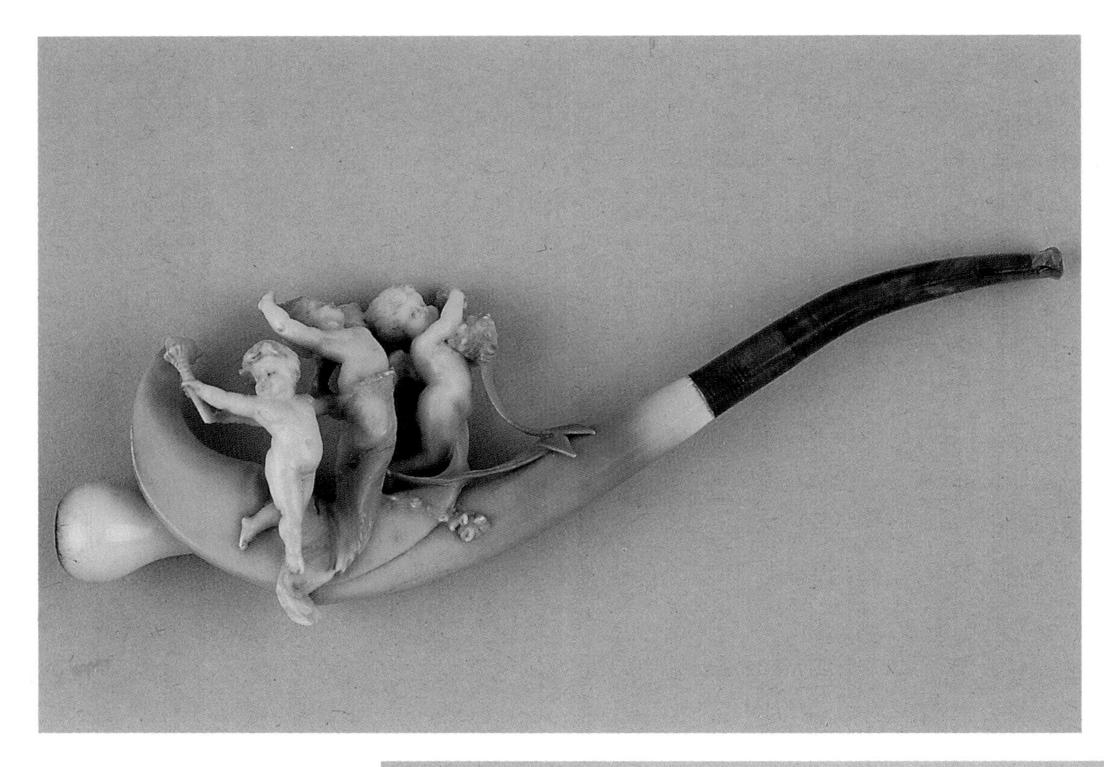

Cheroot holder, three beribboned cherubim cavorting with the Man in the Moon, 6.75" l., 3" h., Bailey, Banks & Biddle, Philadelphia, ca. 1890. *Courtesy of the SP Collection.*

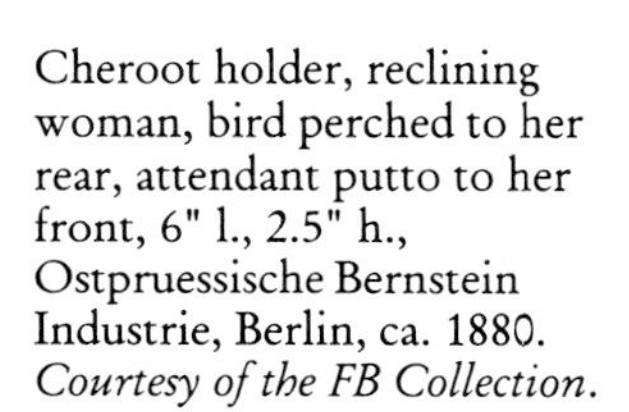

Cheroot holder, reclining woman, bird perched to her rear, attendant putto to her front, 6" l., 2.5" h., Ostpruessische Bernstein Industrie, Berlin, ca. 1880. *Courtesy of the FB Collection.*

Cheroot holder, cherub astride swan, 5" l., 2" h., prob. English, ca. 1880. *Courtesy of the GLD Collection.*

Cheroot holder, cherub seated astride imaginative cetacean mammal, 7" l., 3" h., prob. French, ca. 1890. *Courtesy of a WC Collector.*

Pipe, two cherubs, each astride imaginative cetacean mammal, conch surmounted on head, 9.5" l., 4.75" h., prob. French, ca. 1880. *Courtesy of a WC Collector.*

Table pipe, cherub mounted on marble pedestal, flexible hose, 5.75" h., prob. English, ca. 1890. *Courtesy of the JTB Collection.*

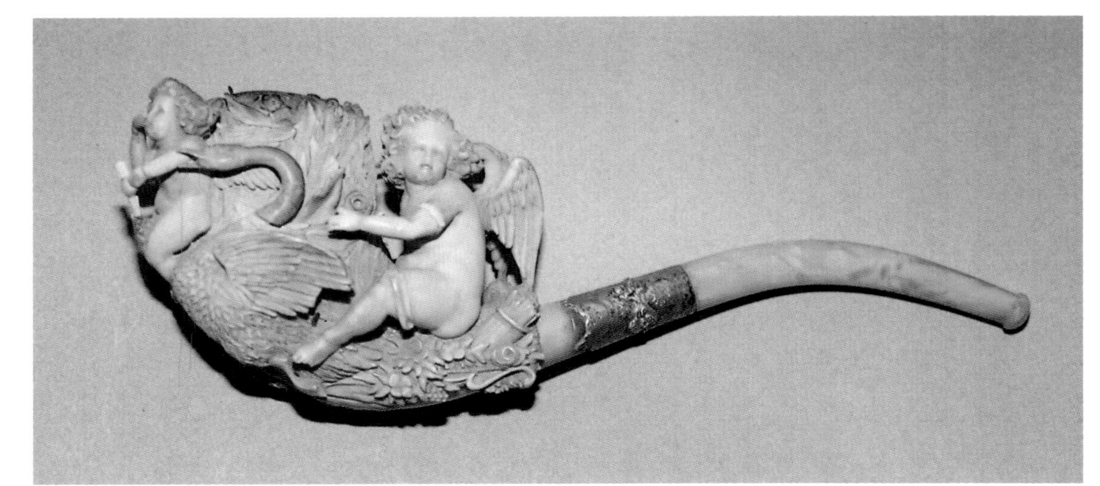

Pipe, cherub in front of bowl, and to the rear, cupid with quiver of arrows, swans juxtaposed on either side of the bowl, decorated silver mount, 10.5" l., 3.5" h., prob. English, ca. 1875. *Courtesy of the JTB Collection.*

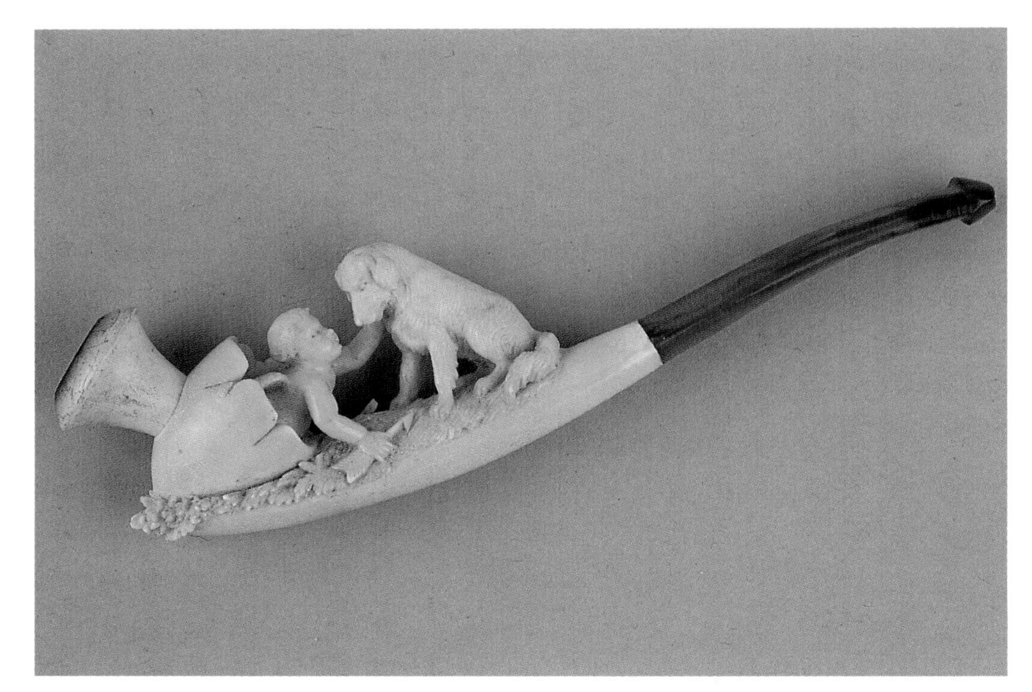

Cheroot holder, cupid emerging from egg shell to greet dog, 5.5" l., 1.5" h., prob. American or English, ca. 1890. *Courtesy of the AZ Collection.*

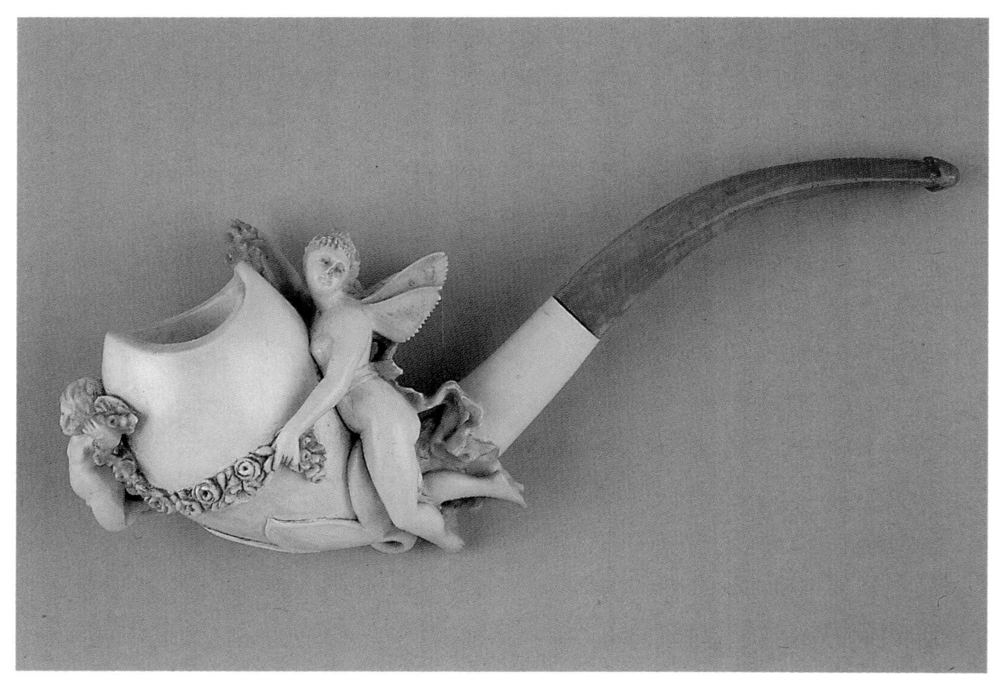

Pipe, oread juxtaposed to cherub, intertwined by garlands, 7.5" l., 3" h., prob. English or American, ca. 1890. *Courtesy of the SP Collection.*

Cheroot holder, escutcheon, with crown surmounted, bordered by two fairies, those imaginary supernatural beings, 7" l., 3.25" h., prob. German, ca. 1880. *Courtesy of the GC Collection.*

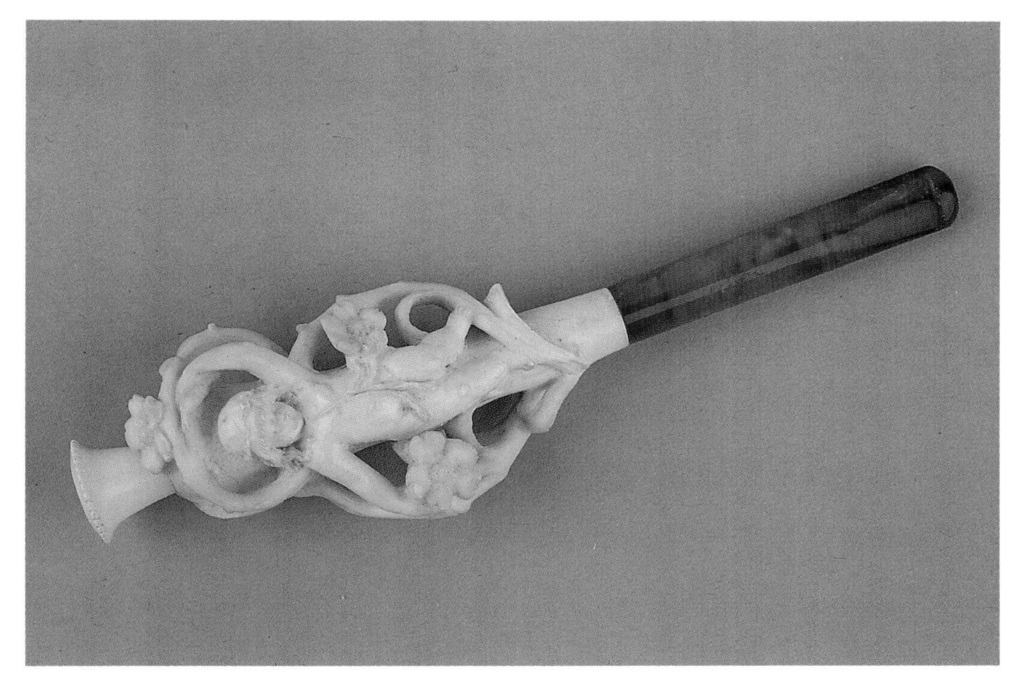

Cigar holder, young girl circumscribed by vines and flower petals, 6" l., prob. Austrian-Hungarian, ca. 1880. *Courtesy of the RHW Collection.*

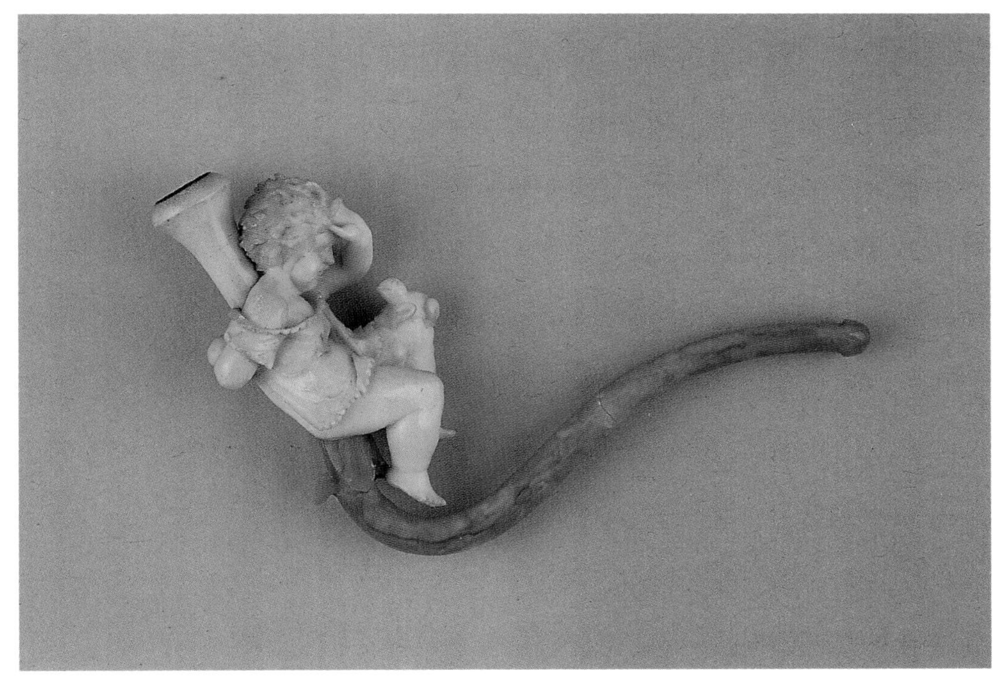

Cheroot holder, young child fending off frog lusting for her apple, 6.25" l., 4.25" h., prob. German, ca. 1890. *Courtesy of the SP Collection.*

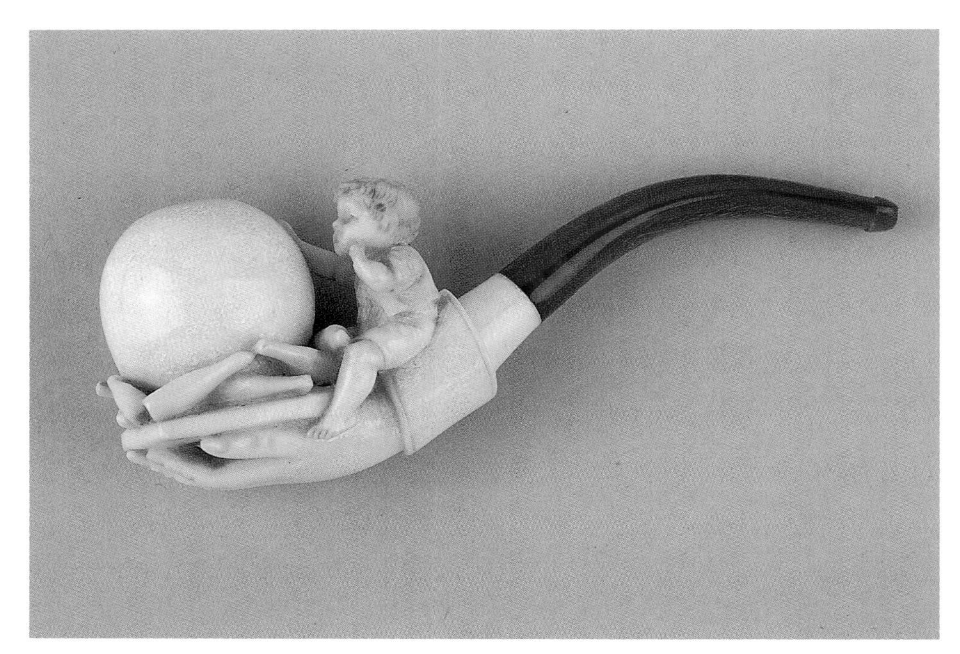

Pipe, young lad in serious study of larger-than-life bowling ball and ten pins, 6.5" l., 2.5" h., prob. American or English, ca. 1890. *Courtesy of the Author's Collection.*

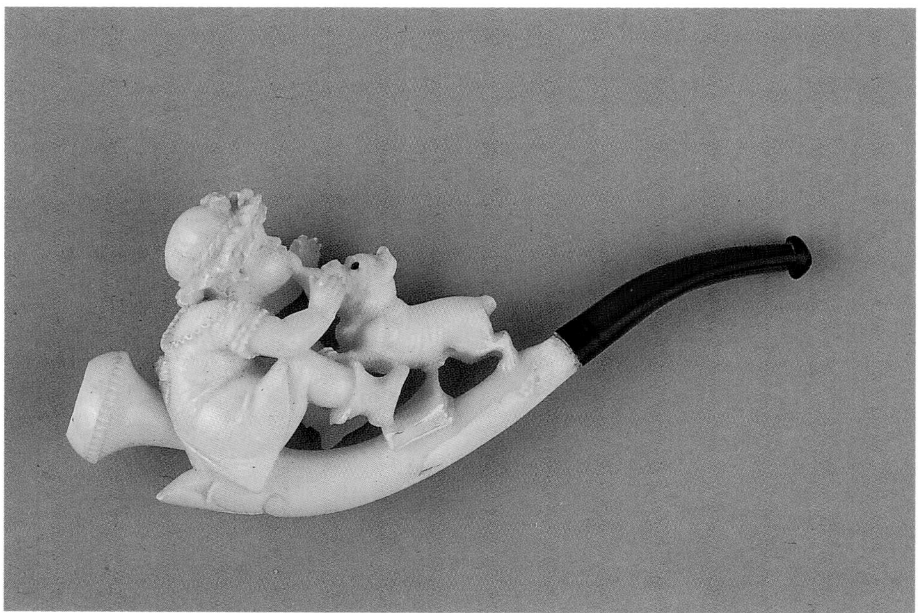

Cheroot holder, young girl bussing dog, 5" l., 2" h., prob. American or English, ca. 1890. *Courtesy of the Author's Collection.*

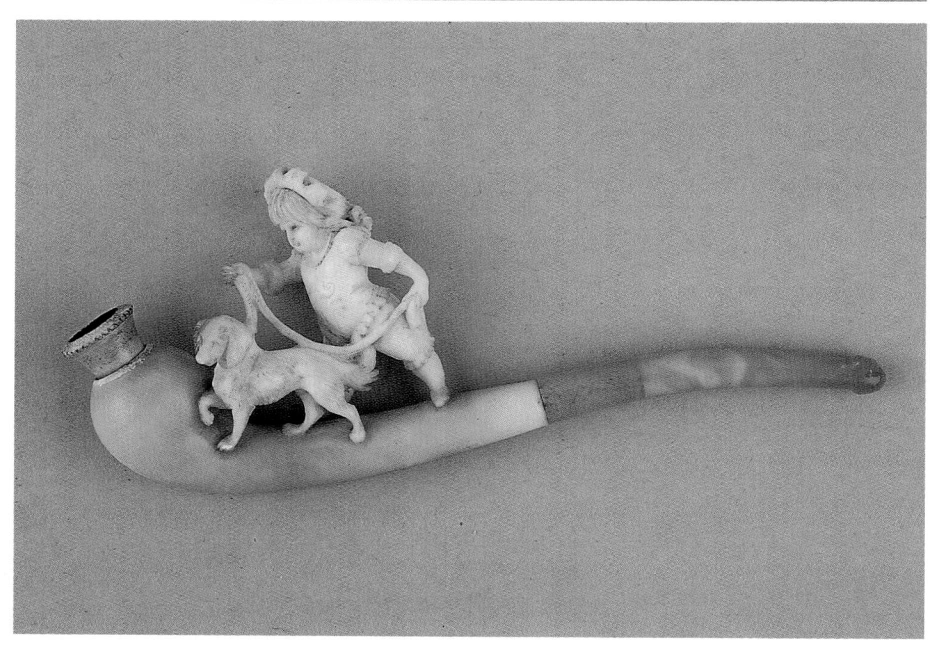

Cheroot holder, young girl walking dog, 6" l., 2.5" h., prob. American or English, ca. 1890. *Courtesy of the Author's Collection.*

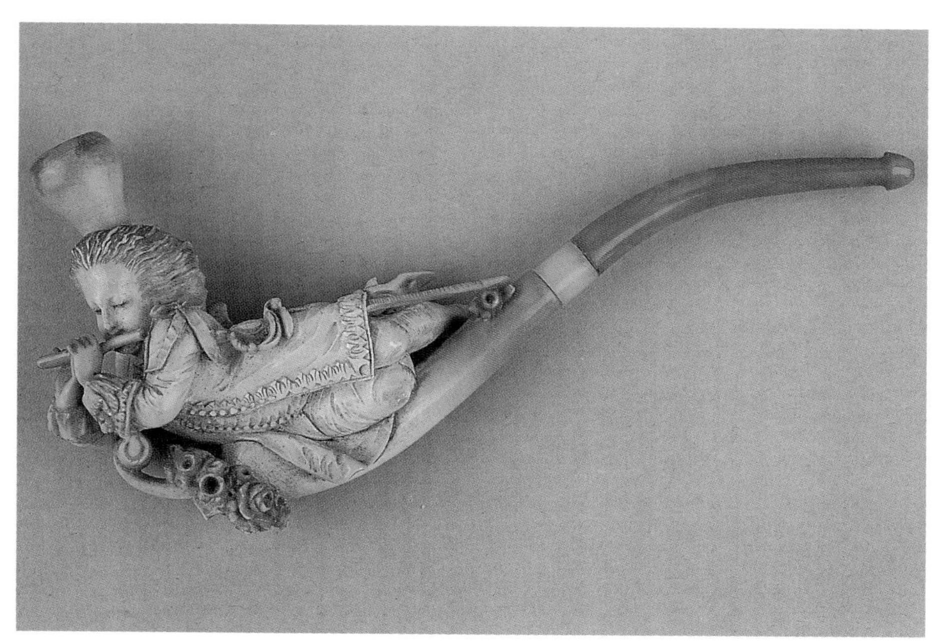

Cheroot holder, young flutist in medieval garb, 6.5" l., 3" h., Ludwig Hartmann & Eidam, Wien, ca. 1880. *Courtesy of the Author's Collection.*

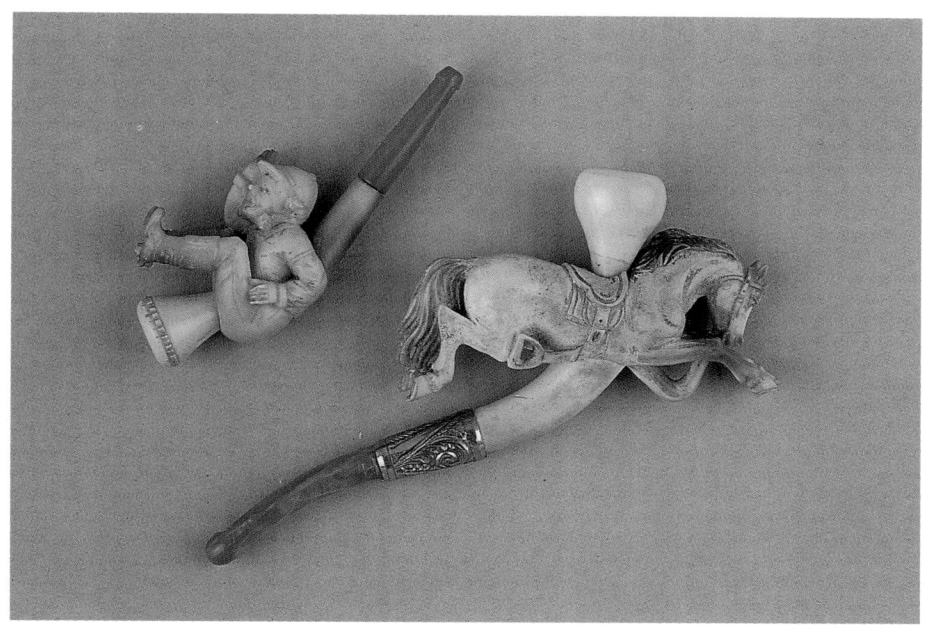

Cheroot holders, cased set, young jockey, 2.5" l., and horse, 3.5" l., prob. English, ca. 1890. *Courtesy of the SP Collection.*

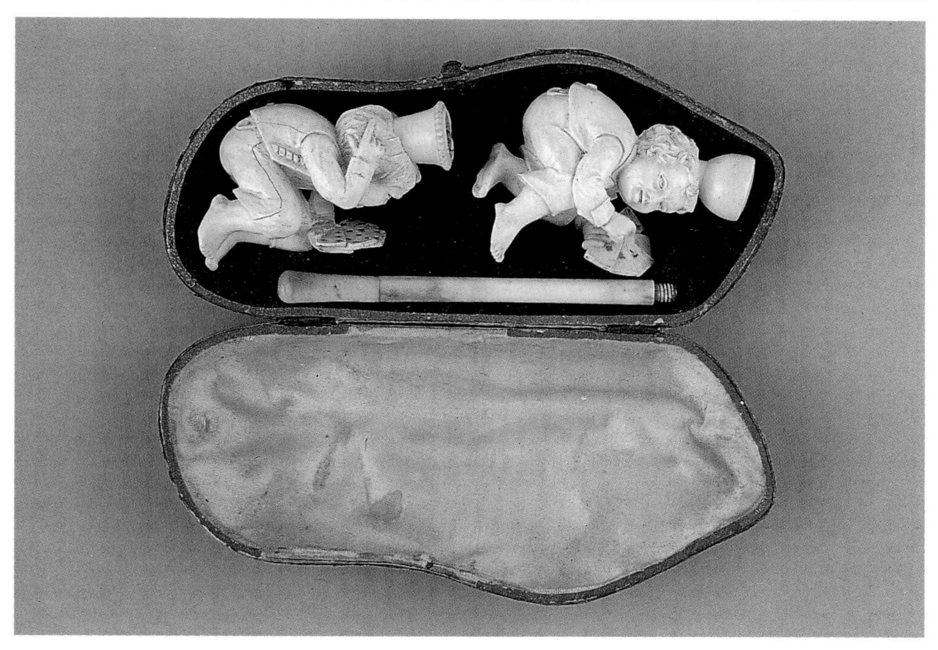

Cheroot holders, cased set, two boys playing cards, one obviously winning, the other losing, each 2" l., 2.5" h., ivory push stem, prob. English, ca. 1890. *Courtesy of the SP Collection.*

Cheroot holders, uncased.

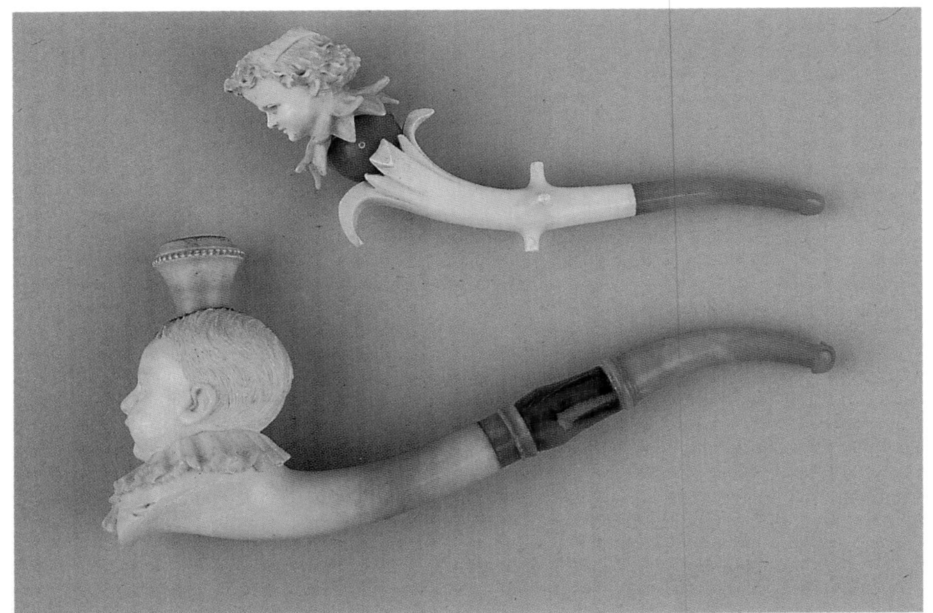

Holders, cigarette holder, figurative head of young child, 5.5" l., 2" h., case stamped A. E. Kettle, London, E. C., case stamped "Crystal Palace, 1884" (top); cheroot holder, figurative head of young child, 7" l., 3" h., American, ca. 1900 (bottom). (It is rare when a meerschaum tells a real-life story; this one does. A carver in New York City was commissioned to render this piece using a portrait given to him by a family wanting to have a memento celebrating the second birthday of the first son.) *Courtesy of the Author's Collection.*

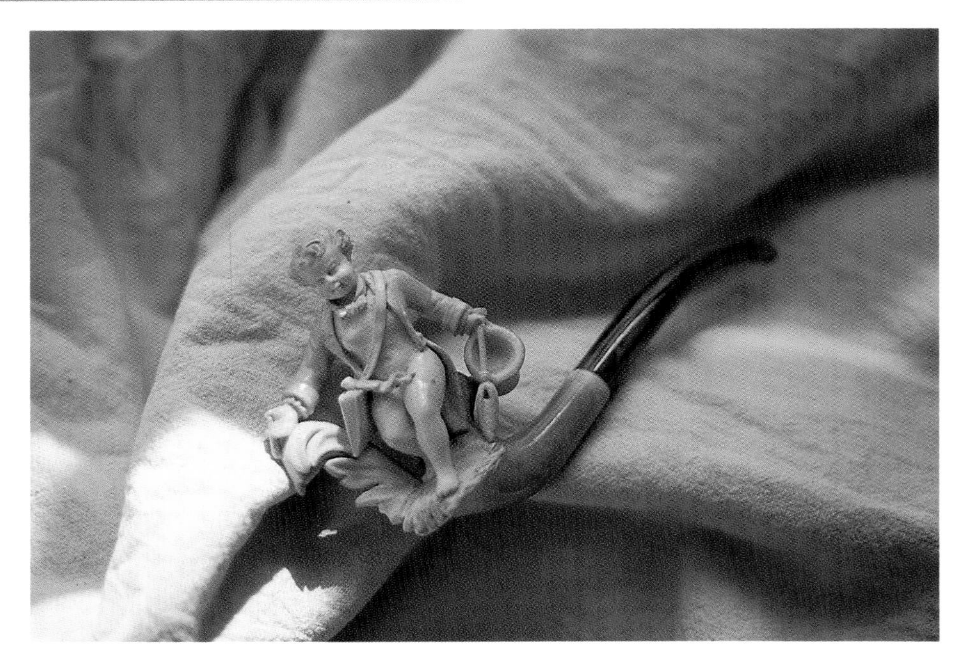

Cheroot holder, Puck, after the mischievous sprite from Shakespeare's *A Midsummer Night's Dream*, 7" l., 3.5" h., prob. American, ca. 1890. *Courtesy of the MG Collection.*

Cheroot holder, three girls in bathing suits exiting the surf, 5.75" l., 2" h., prob. German, ca. 1890. *Courtesy of the AZ Collection.*

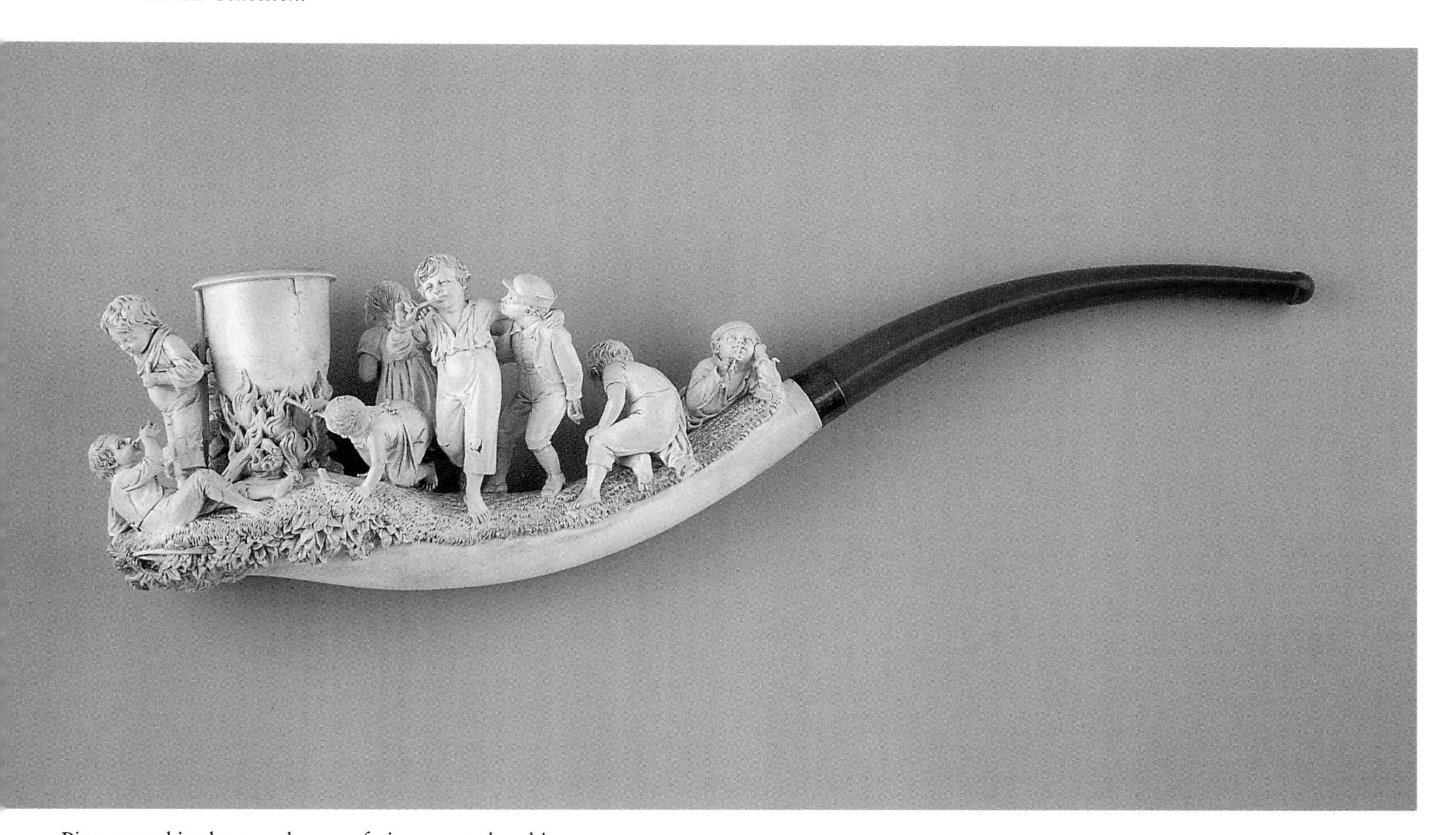

Pipe, carved-in-the-round scene of nine assorted urchins experimenting with "The First Smoke" and the aftereffects, 17.25" l., 4.5" h., Carl Hiess, Wien, ca. 1875. (The fitted case's clasp is engraved "K.K. Priv" [Kaiserlich-Königlich, translated as imperial and royal]; after 1868, this was a trademark identified with the Austrian Emperor and King of Hungary, Franz Josef I.) *Courtesy of the SP Collection.*

Chapter XIV. Whimsy, Fancy, Quaint, and Curious

In the course of collecting, I have occasionally given thought that some carvers must have had a very strange sense of humor, and that they also must have had some leisure time at the bench and liberal use of raw materials because I have seen—and own—many meerschaums that are quite impractical smoking utensils that could not have been the serious output of anyone trying to earn a living in this cottage industry. I classify the too long, too short, too high, and the too bizarre shape or style, all those beyond the pale of pipe-reasonableness, as oddities. Impractical to smoke, perhaps, but some, however outlandish in their appearance, are handsome additions to a collection. For example, the limited assortment of those illustrated in this chapter might have been produced to engender light conversation, maybe even hearty laughter among serious smokers. Nonetheless, they are deserving of being shown, and I devote a chapter to their cause.

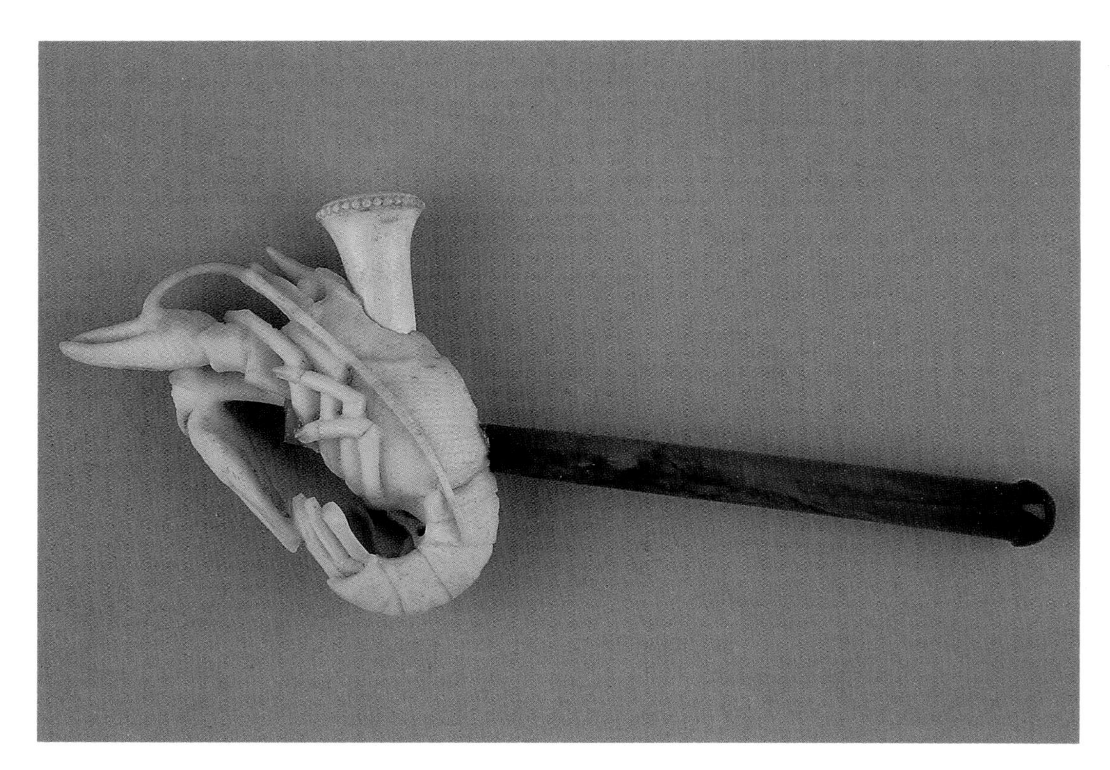

Cheroot holder, figurative of crayfish with amber push stem, 5" l., M. Saury, Constantinople, ca. 1900. *Courtesy of the SP Collection.*

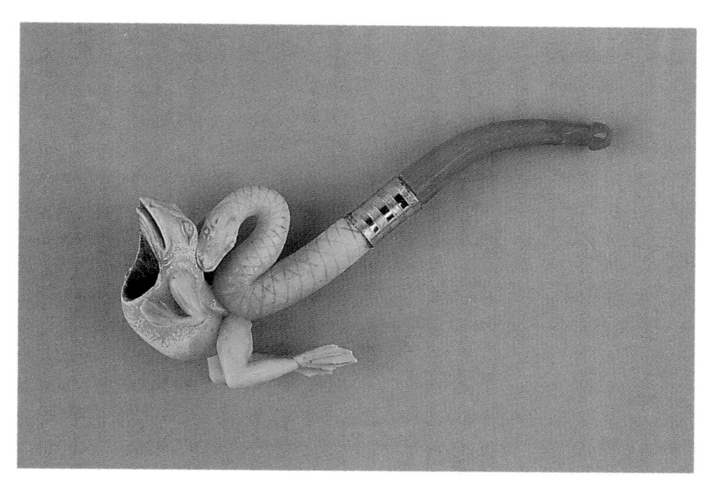

Pipe, diminutive, figurative of snake biting frog, 4.5" l., prob. French, ca. 1900. *Courtesy of the SP Collection.*

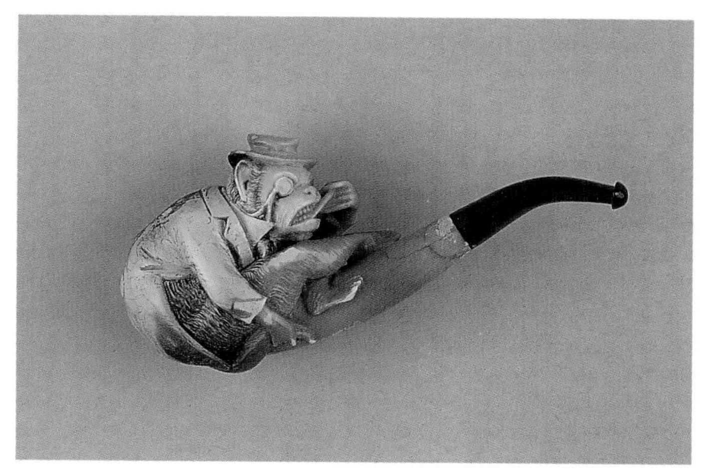

Cheroot holder, figurative of monkey, sporting crushed top hat, tails, and monocle, smoking a cigar, 5" l., 2.75" h., prob. German, ca. 1890. *Courtesy of the SP Collection.*

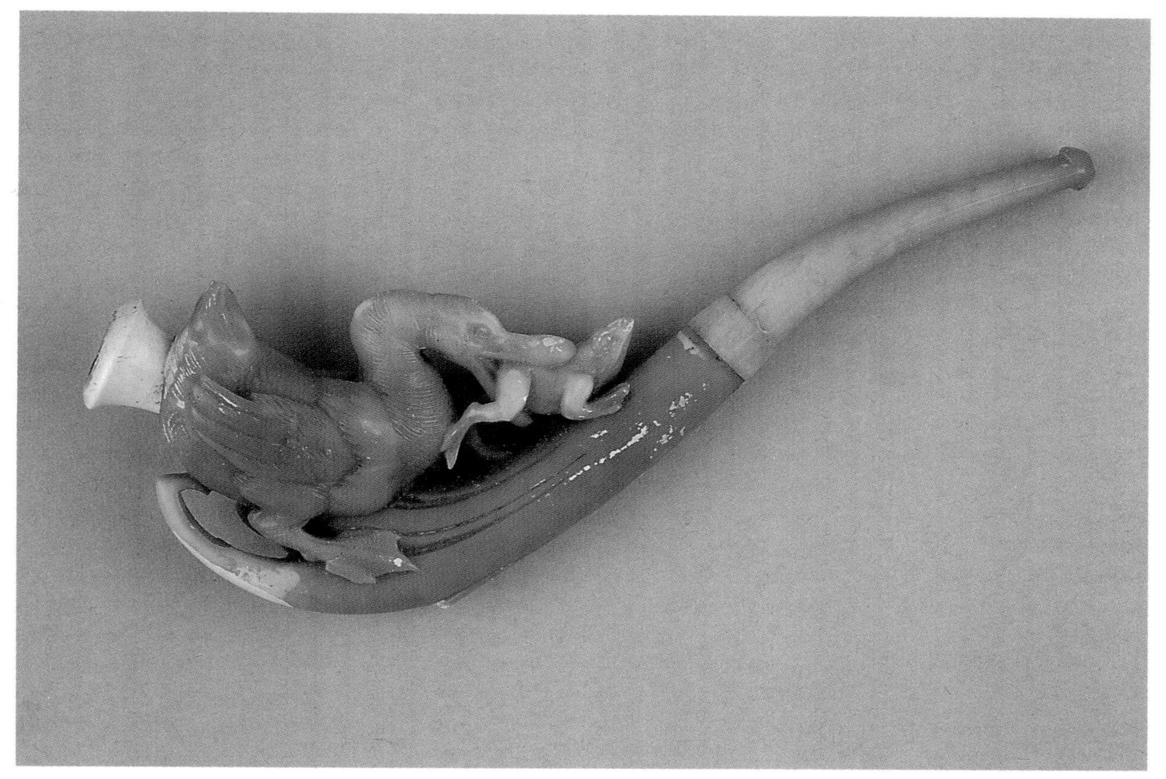

Cheroot holder, goose swallowing frog, 6.5" l, 2.75" h., Au Pacha, Paris, ca. 1890. *Courtesy of the SP Collection.*

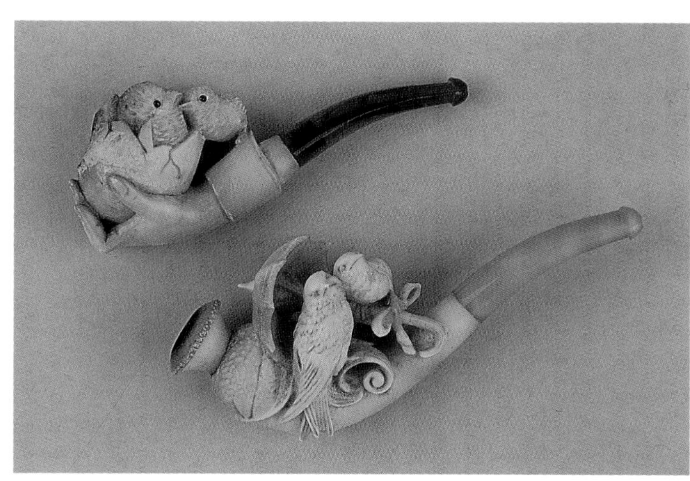

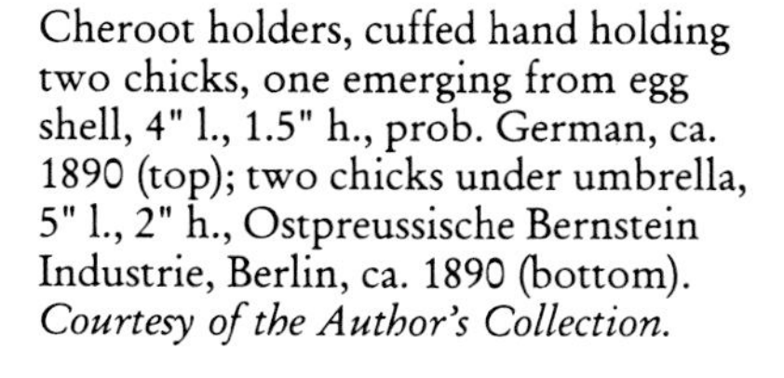

Cheroot holders, cuffed hand holding two chicks, one emerging from egg shell, 4" l., 1.5" h., prob. German, ca. 1890 (top); two chicks under umbrella, 5" l., 2" h., Ostpreussische Bernstein Industrie, Berlin, ca. 1890 (bottom). *Courtesy of the Author's Collection.*

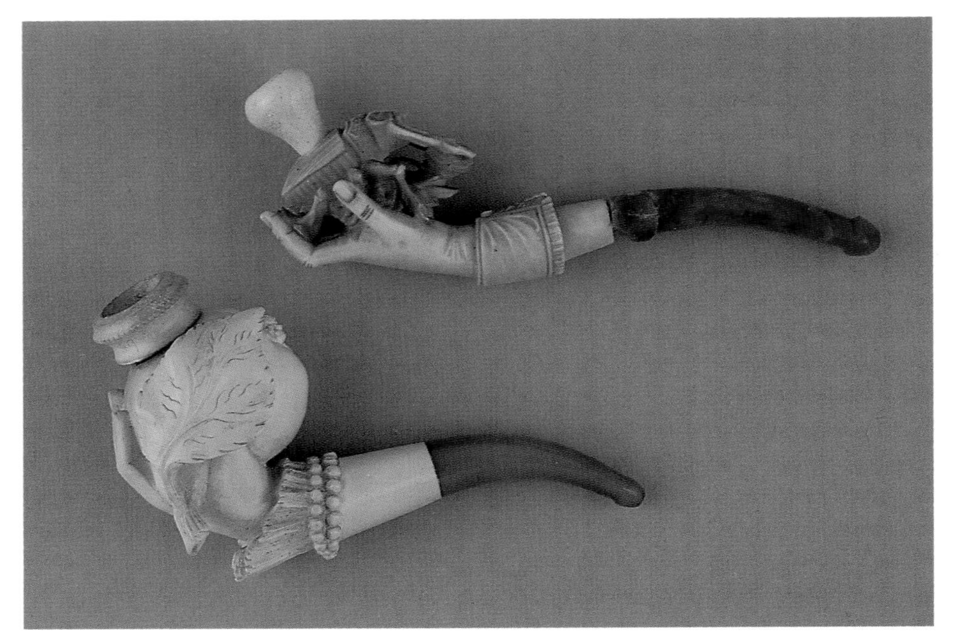

Holders, cigarette holder, lady's cuffed hand holding dung beetle, 4.5" l., 2" h., Ludwig Hartmann & Eidam, Wien, ca. 1880 (top); cheroot holder, lady's lace-cuffed hand holding pomegranate, 4" l., 2.5" h., prob. German, case stamped "5/12/89" (bottom). *Courtesy of the Author's Collection.*

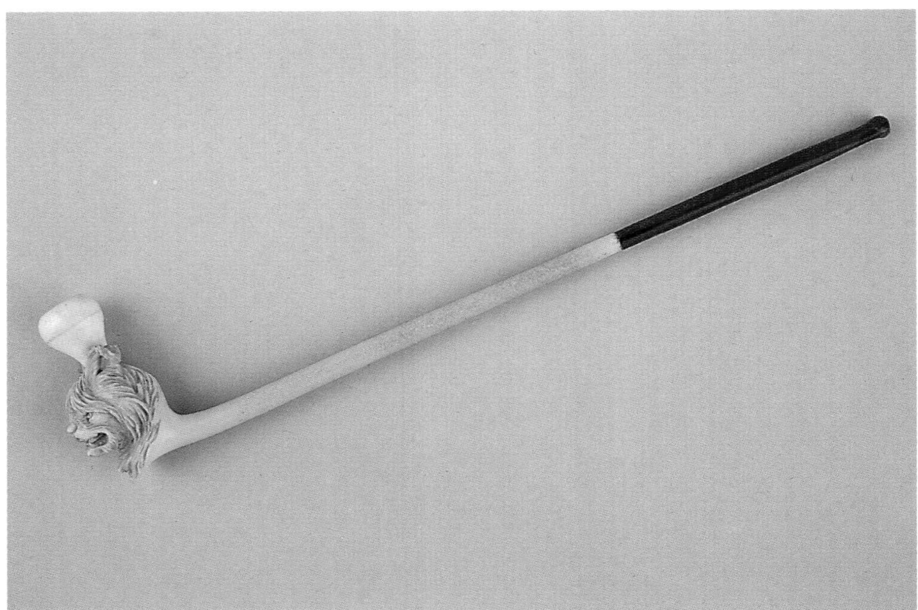

Cigarette holder, figurative head of wild dog, 9" l., 2" h., Ludwig Hartmann & Eidam, Wien, ca. 1900. *Courtesy of the Author's Collection.*

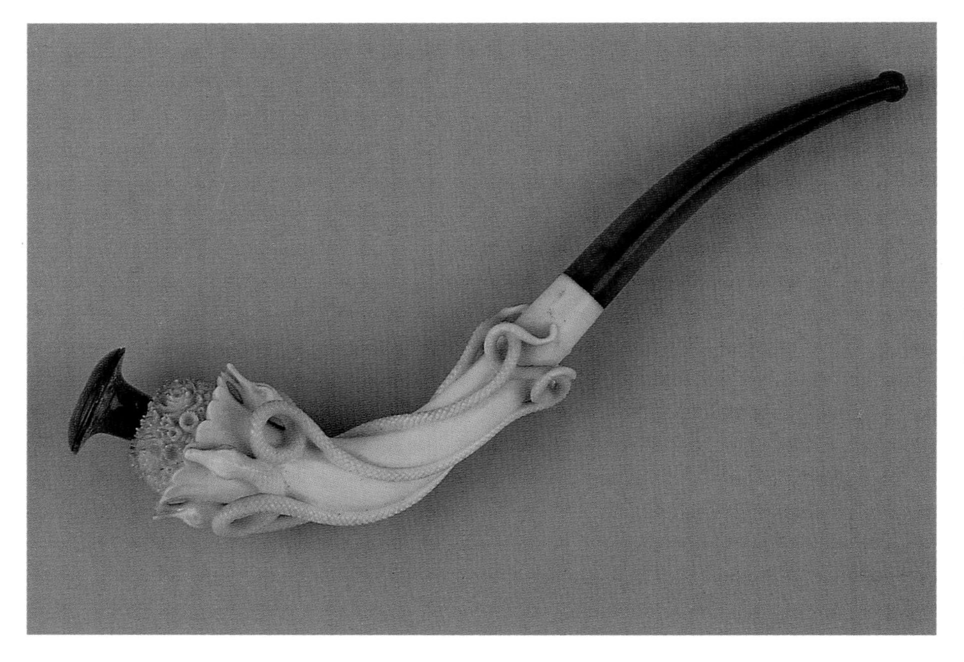

Cheroot holder, four spiraling serpents circumscribing shank, 9" l., 3" h., amber coloring bowl accent, WDC, American, ca. 1900. *Courtesy of the Author's Collection.b*

Cheroot holders, Sebastopol-style, terminated in figurative head of wild boar, 3.5" l., 5.5" h., wood push stem, Jos. Dolezal, Prag, ca. 1890 (top); Sebastopol-style terminated in figurative head of horse, 3" l., 5" h., Carl Hiess, Wien, ca. 1900 (bottom). *Courtesy of the Author's Collection.*

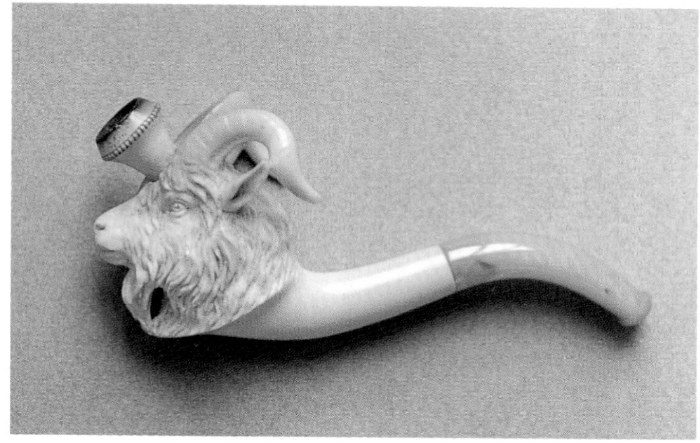

Cheroot holder, figurative head of ram, 5.25" l., 2.25" h., prob. English, ca. 1885. *Courtesy of the JTB Collection.*

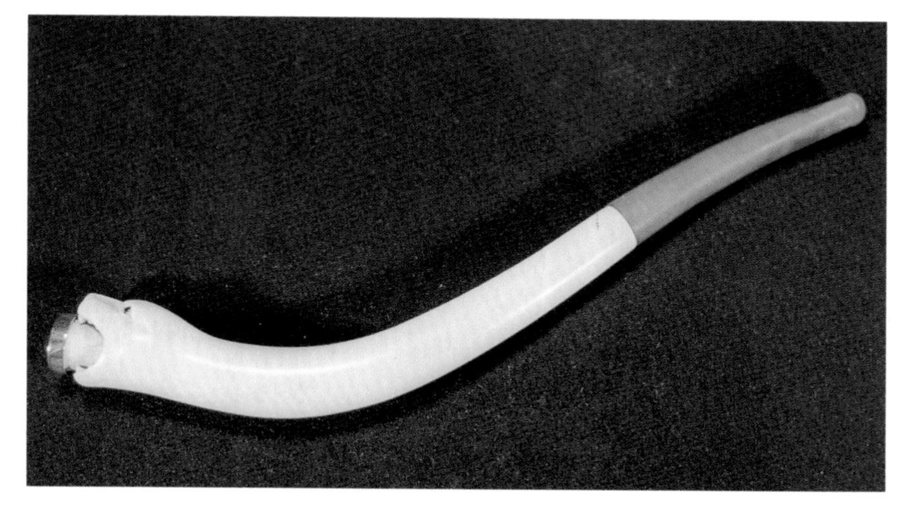

Cigarette holder, terminated in figurative head of seal, 9.5" l., 1" h., Emil Pietzsch, Dresden, ca. 1900. *Courtesy of a WC Collector.*

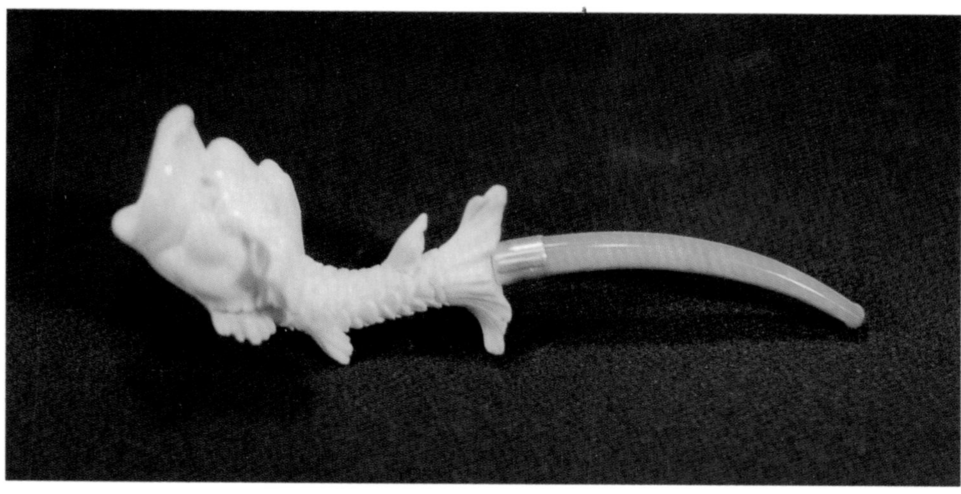

Pipe, expressed as imaginative cetacean mammal, 7" l., 2" h., prob. French, ca. 1900. *Courtesy of a WC Collector.*

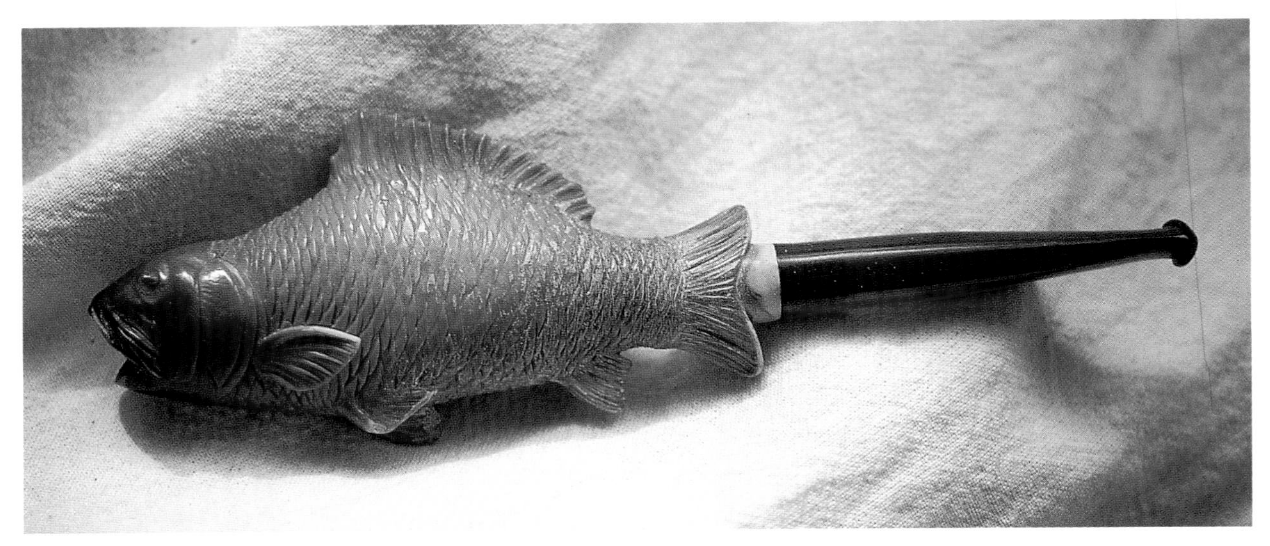

Cigarette holder, fish, red amber push stem, 6" l., 2" h., prob. Austrian, ca. 1900. *Courtesy of the MG Collection.*

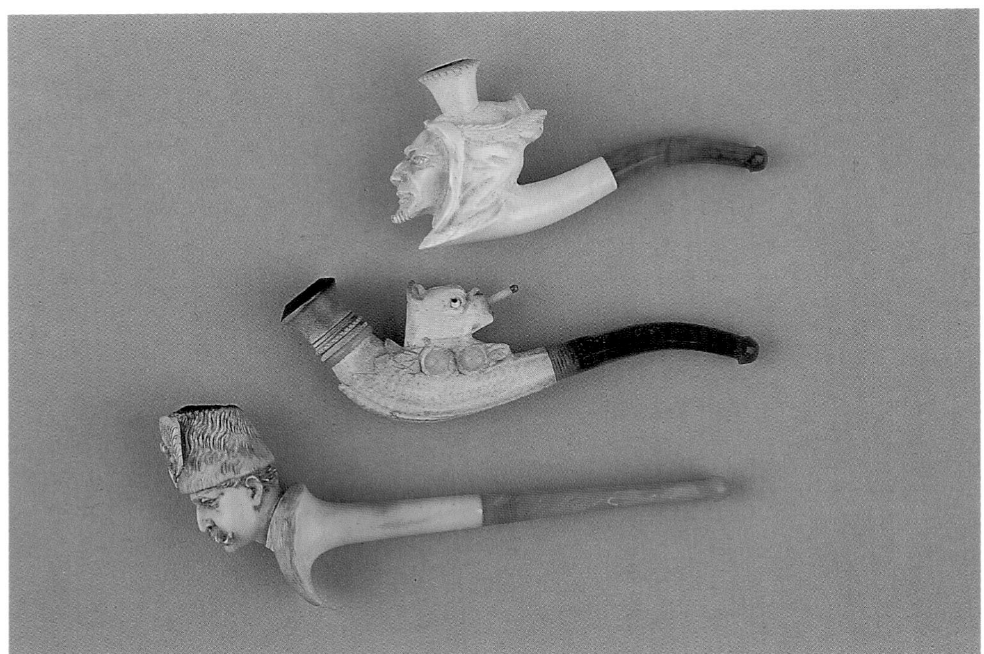

Cigarette holders, figurative head of Mephistopheles, 3" l., 1.5" h., prob. French, ca. 1900 (top); surmounted figurative head of dog smoking cigarette, 4" l., 1.5" h., prob. German, ca. 1900 (center); figurative head of Russian cossack, 4.5" l., 2" h., J.F. Paris, ca. 1900 (bottom). *Courtesy of the Author's Collection.*

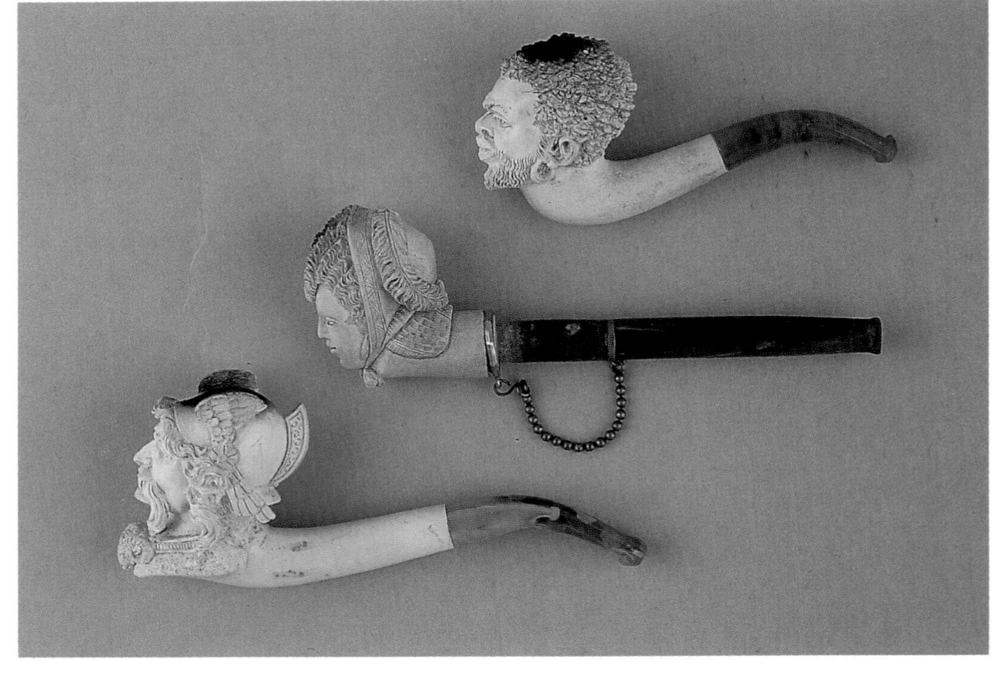

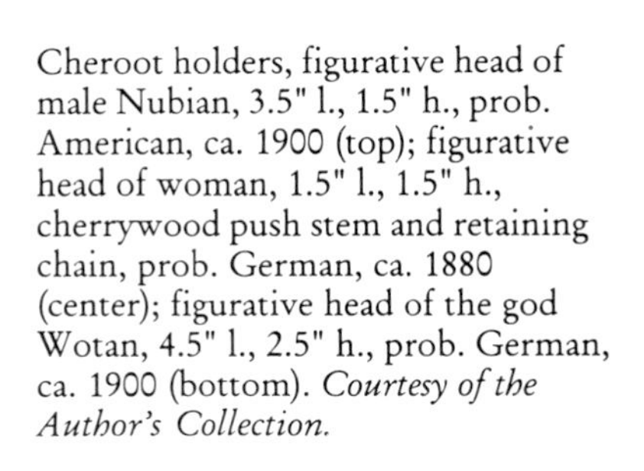

Cheroot holders, figurative head of male Nubian, 3.5" l., 1.5" h., prob. American, ca. 1900 (top); figurative head of woman, 1.5" l., 1.5" h., cherrywood push stem and retaining chain, prob. German, ca. 1880 (center); figurative head of the god Wotan, 4.5" l., 2.5" h., prob. German, ca. 1900 (bottom). *Courtesy of the Author's Collection.*

Cheroot holder, dragon surmounted, forked tail circumscribing shank, 5.25" l., 2" h., prob. English, ca. 1890. *Courtesy of the JTB Collection.*

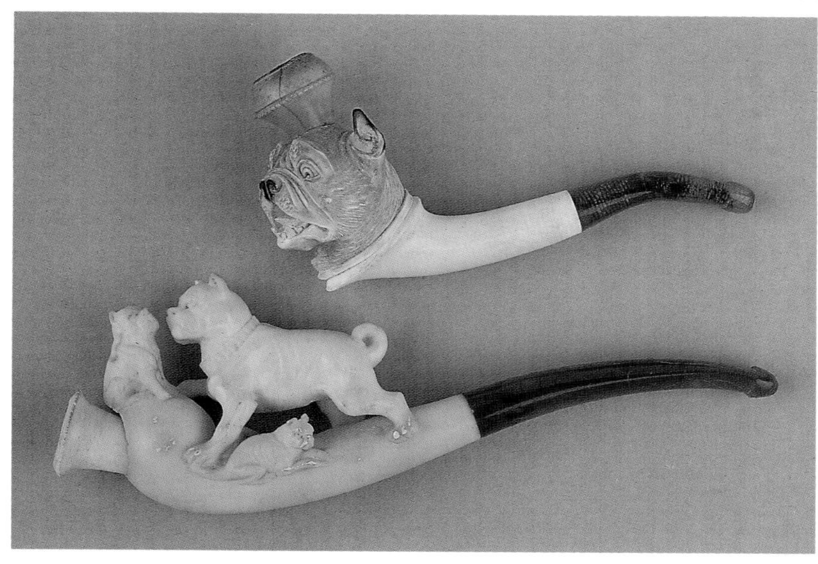

Cheroot holders, figurative head of pug, 4.5" l., 2" h., prob. German, ca. 1890 (top); boxer with cat astride shank, 6" l., 2" h., prob. French, ca. 1890 (bottom). *Courtesy of the Author's Collection.*

Cheroot holder, two pugs at play on a piano with winged dragon accent, 5" l., 2" h., J.G. Gärtner, Dresden, ca. 1900. *Courtesy of the Author's Collection.*

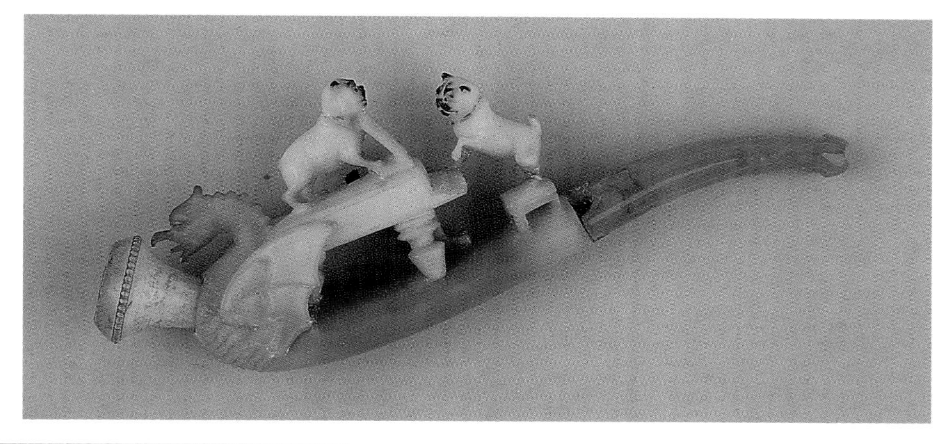

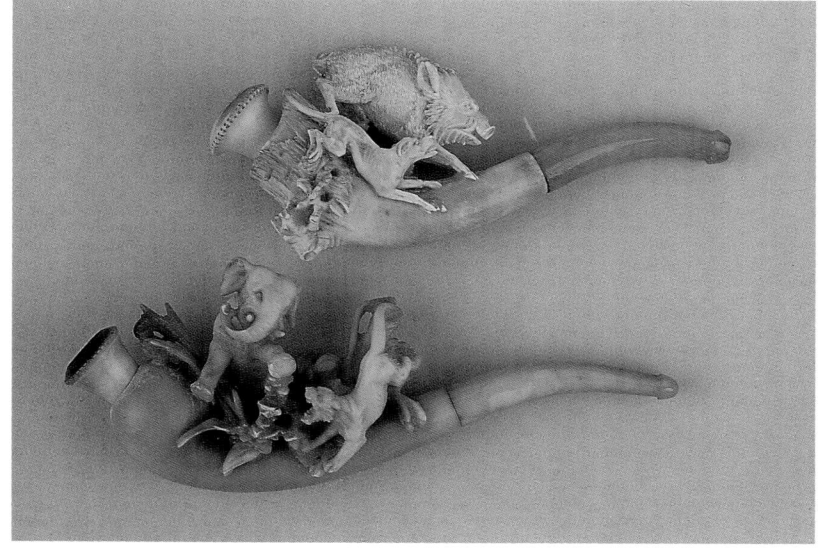

Cheroot holders, wild boar and dog astride, 5" l., 2.5" h., prob. German, ca. 1880 (top); figurative of elephant and cougar astride, 6" l., 2.5" h., Salmon & Gluckstein, London, ca. 1880 (bottom). *Courtesy of the Author's Collection.*

Cheroot holder, figurative head of wild boar with inset glass eyes, cherrywood push stem with amber mouthpiece, 6.25" l., 1.75" h., prob. German, ca. 1870. *Courtesy of the GC Collection.*

Cigar holders, matador and toro surmounted, 5" l., 2" h., prob. German, ca. 1890 (top); hound surmounted, 5.5" l., 1.75" h., G. G. Kast Nachfabriek, Stuttgart, ca. 1900 (bottom). *Courtesy of the Author's Collection.*

Cigar holder, great Dane surmounted, 6.5" l., 2" h., prob. German, ca. 1900. *Courtesy of the SP Collection.*

Cigar holder, gentleman in top hat and cutaway accompanied by two children, 6.5" l., Emanuel Czapek, Prag, ca. 1900. *Courtesy of the SP Collection.*

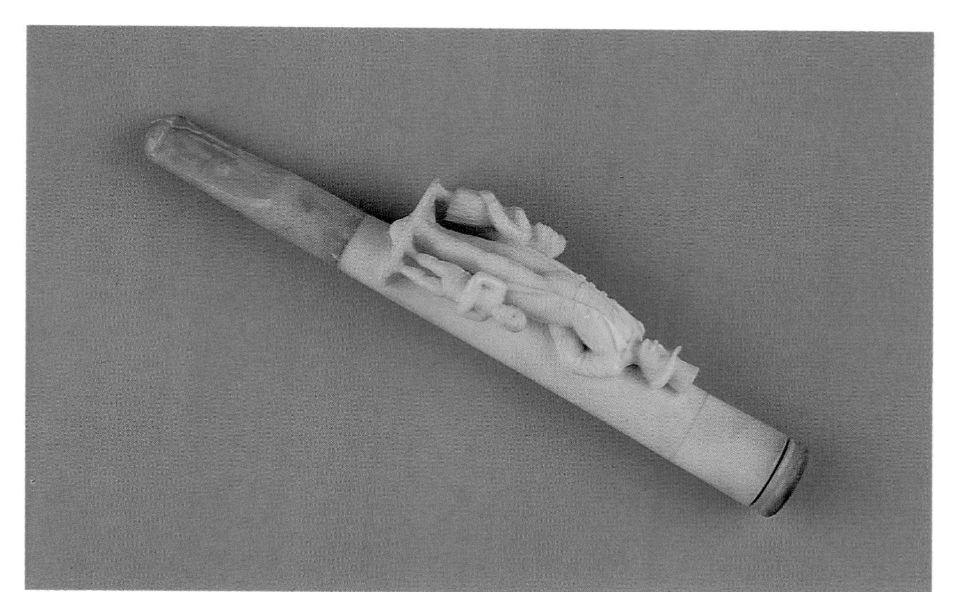

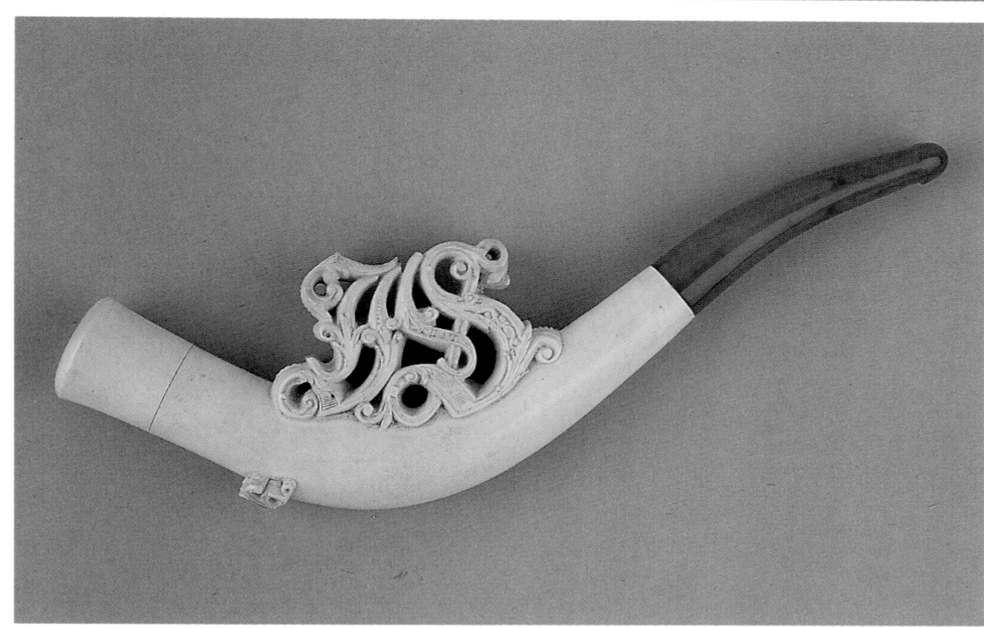

Cheroot holder, scrolled monogram surmounted, "1886" on front, 7.5" l., 2" h., prob. German. *Courtesy of the RHW Collection.*

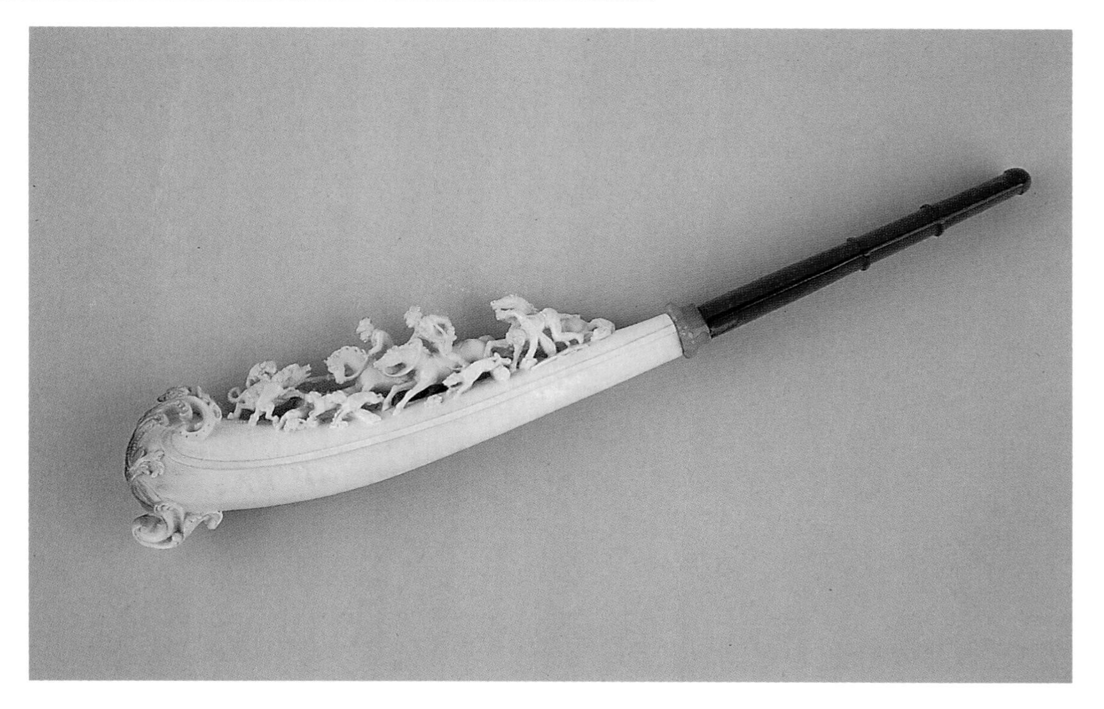

Cigar holder, Tatar foresters on boar hunt, 11" l., 2" h., Austrian-Hungarian, ca. 1900. *Courtesy of the Author's Collection.*

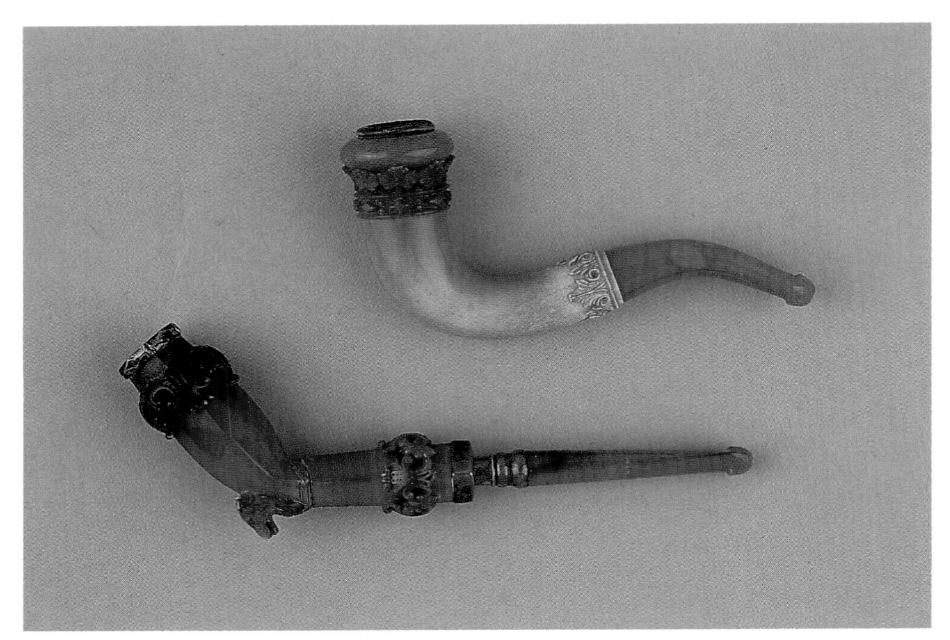

Cheroot holders, amber rim ring, 5" l., 2" h., Au Pacha, Paris, insert paper label "Nov. 1873" (top); two-piece, figurative head of dog on front, silver military bit, 6.5" l., 2" h., prob. German, ca. 1900 (bottom). *Courtesy of the Author's Collection.*

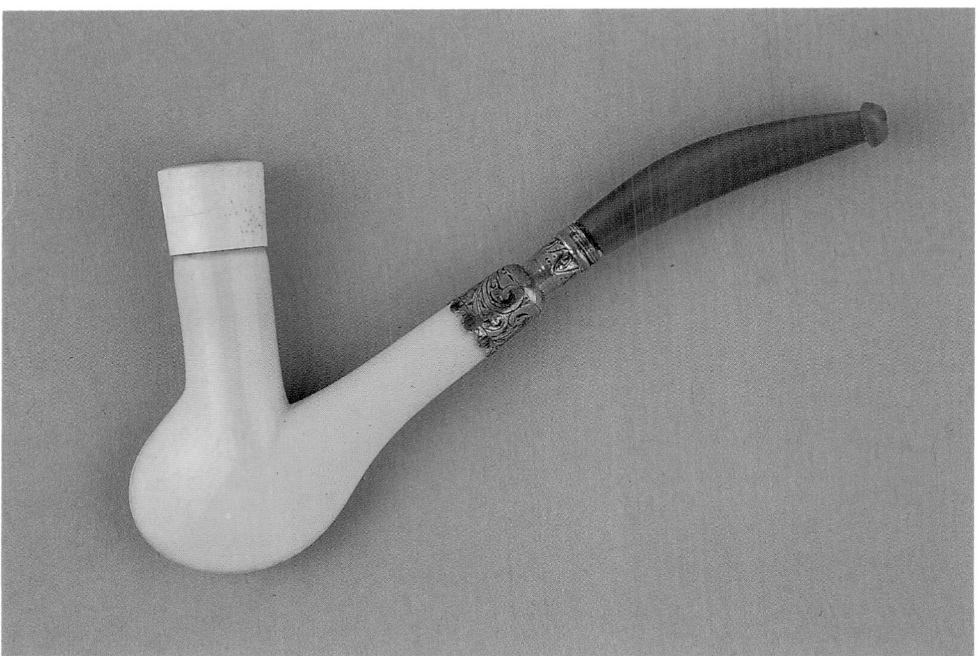

Cheroot holder, Viennese art wassersack, chased silver military bit, 6" l., 2.5" h., Emanuel Czapek, Prag, ca. 1900. *Courtesy of the Author's Collection.*

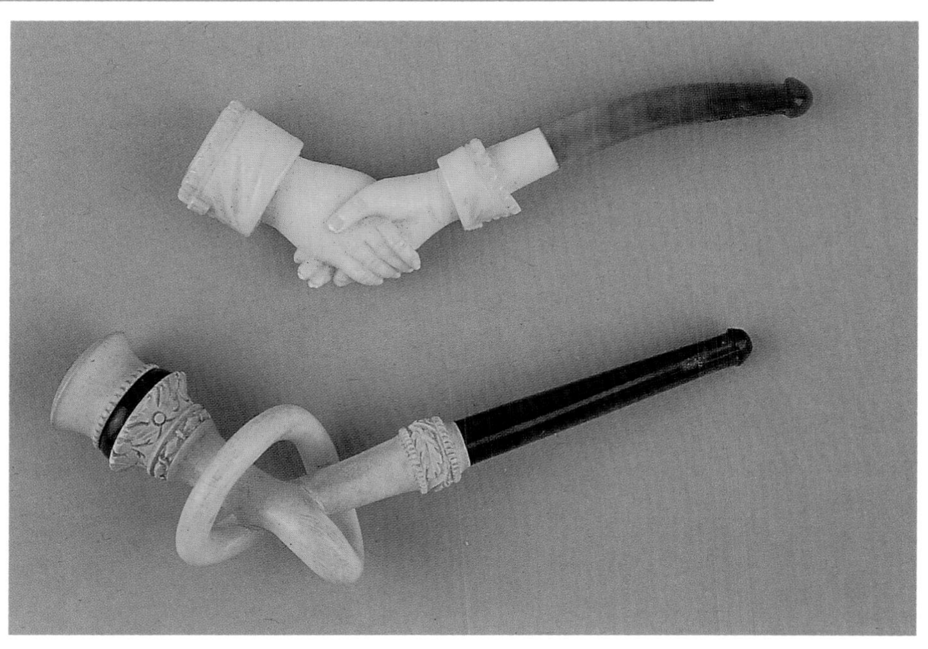

Cheroot holders, clasped hands, symbolic of friendship (or engagement), 5" l., 2" h., prob. German, ca. 1900 (top); spiral knot, amber rim ring, 5" l., 2.5" h., prob. French, ca. 1900 (bottom). *Courtesy of the Author's Collection.*

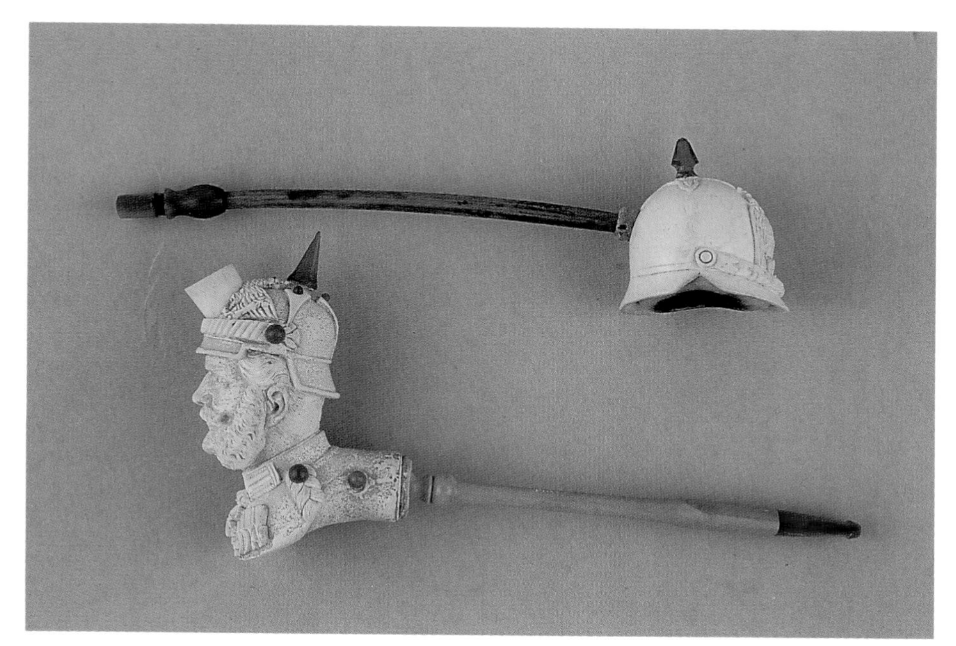

Pipe, pickelhaube helmet, amber finial, 2" h., wood push stem, German, ca. 1875 (top); cheroot holder, figurative head of Prince Karl Otto Furst von Bismarck-Schonhausen, amber bead decor and amber finial on pickelhaube, 2" l., 6.5" h., olive wood push stem, German, ca. 1885 (bottom). *Courtesy of the Author's Collection.*

Cheroot holder, cuffed hand holding stylized bouquet of flowers, 5.5" l., 3" h., prob. French, ca. 1890. *Courtesy of the Author's Collection.*

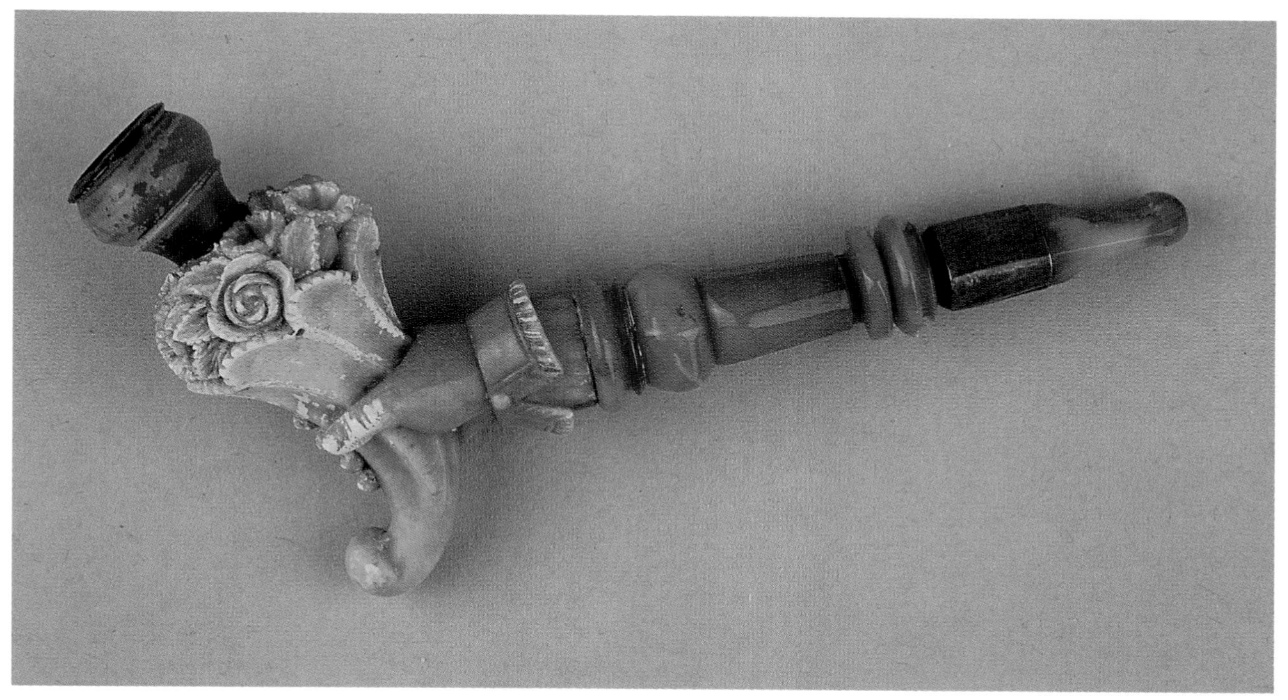

Cheroot holder, cuffed hand holding Derringer-style pistol, 7" l., 3" h., Stehr, New York, ca. 1900. *Courtesy of the MG Collection.*

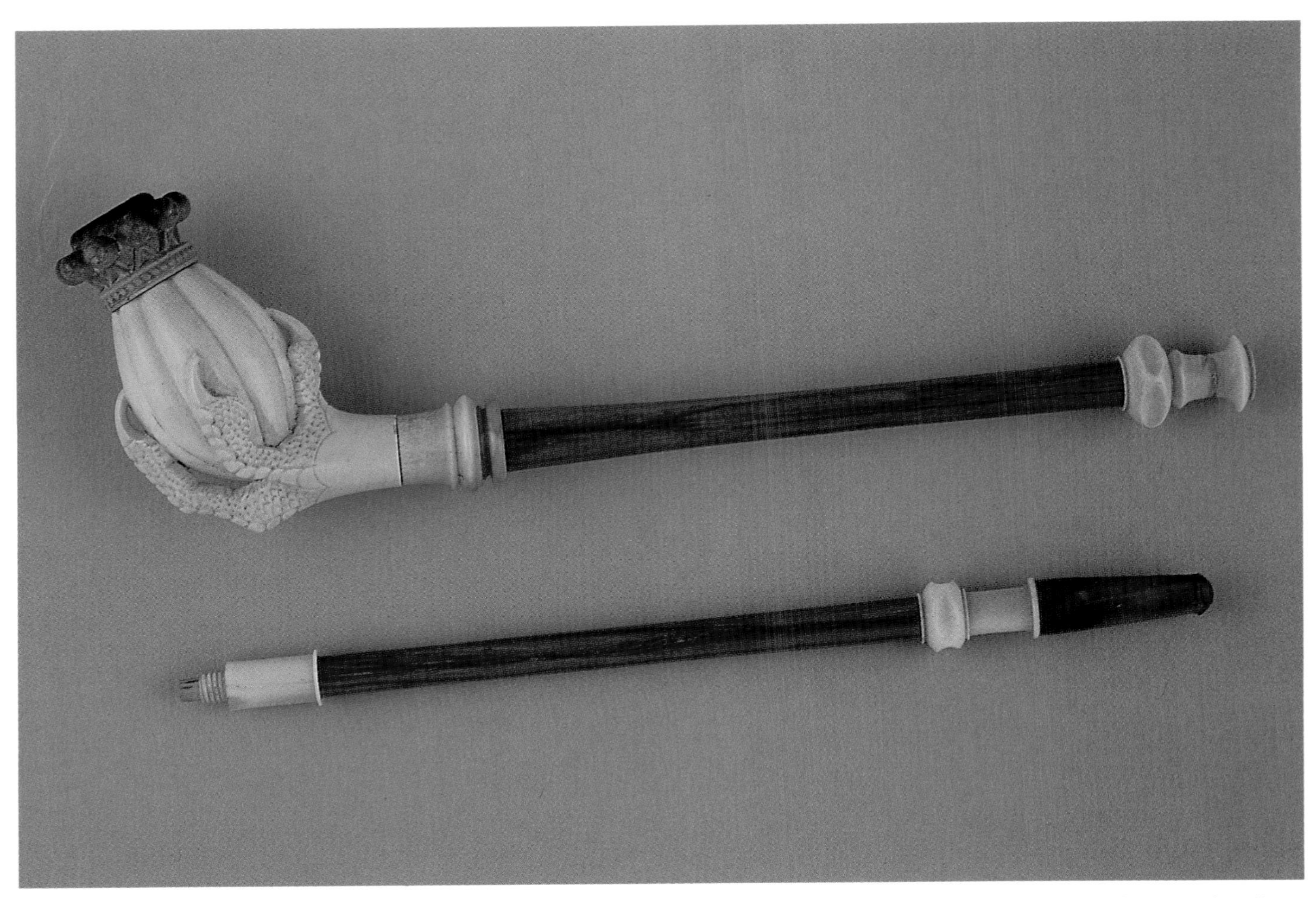

Cheroot holder, talon, mixed media two-piece stem of wood, turned ivory fittings, and amber mouthpiece, 14" l., 2.5" h., German, ca. 1900. *Courtesy of the Author's Collection.*

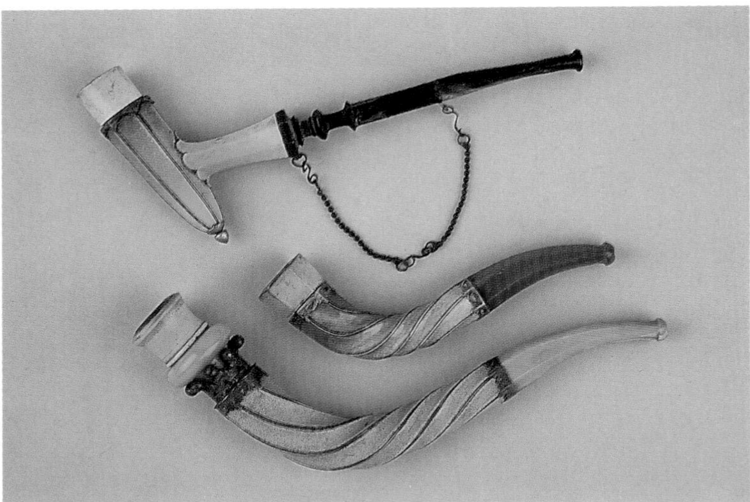

Cheroot holders, Sebastopol-style, inlaid silver wire decor, wood push stem with retaining chain, 6.5" l., 3" h. German, ca. 1900 (top); turned shank, inlaid silver wire decor, 5" l., 2.5" h., German, ca. 1900 (center); turned shank, inlaid silver wire, amber ring below rim, 7.5" l., 3.5" h., German, ca. 1900 (bottom). *Courtesy of the Author's Collection.*

Cheroot holder, Sebastopol-style, reticulated basket-weave overlay on amber, amber accents, amber push stem and retaining chain, 7" l., 5.5" h., German, ca. 1890. *Courtesy of the SP Collection.*

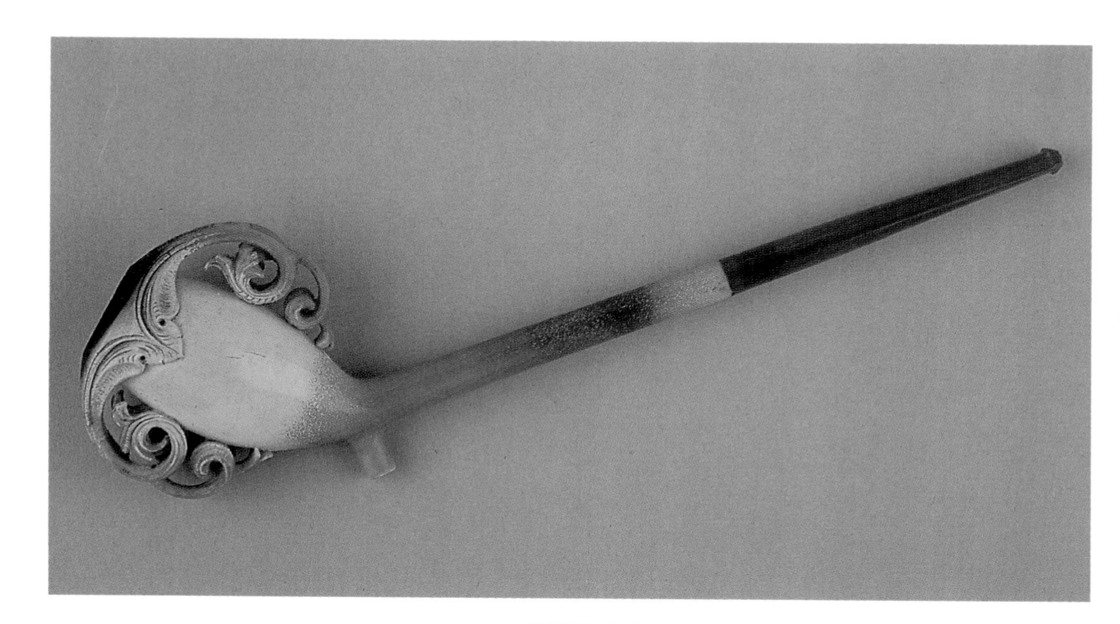

Pipe, art nouveau trim, 9" l., 2.5" h., J. Sommer, Paris, ca. 1910. *Courtesy of the Author's Collection.*

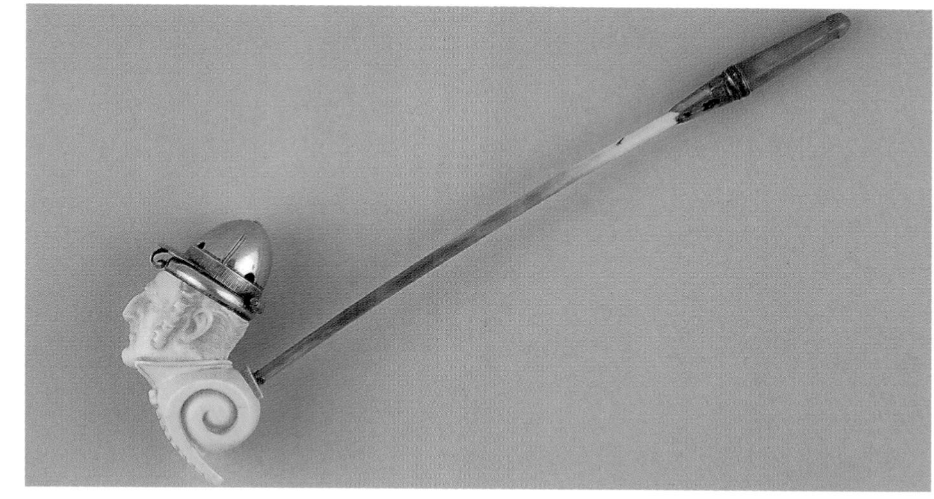

Pipe, figurative head of man as ship's masthead, silver windcap, albatross bone stem, 8.5" l., 2.5" h., prob. French, ca. 1900. *Courtesy of the SP Collection.*

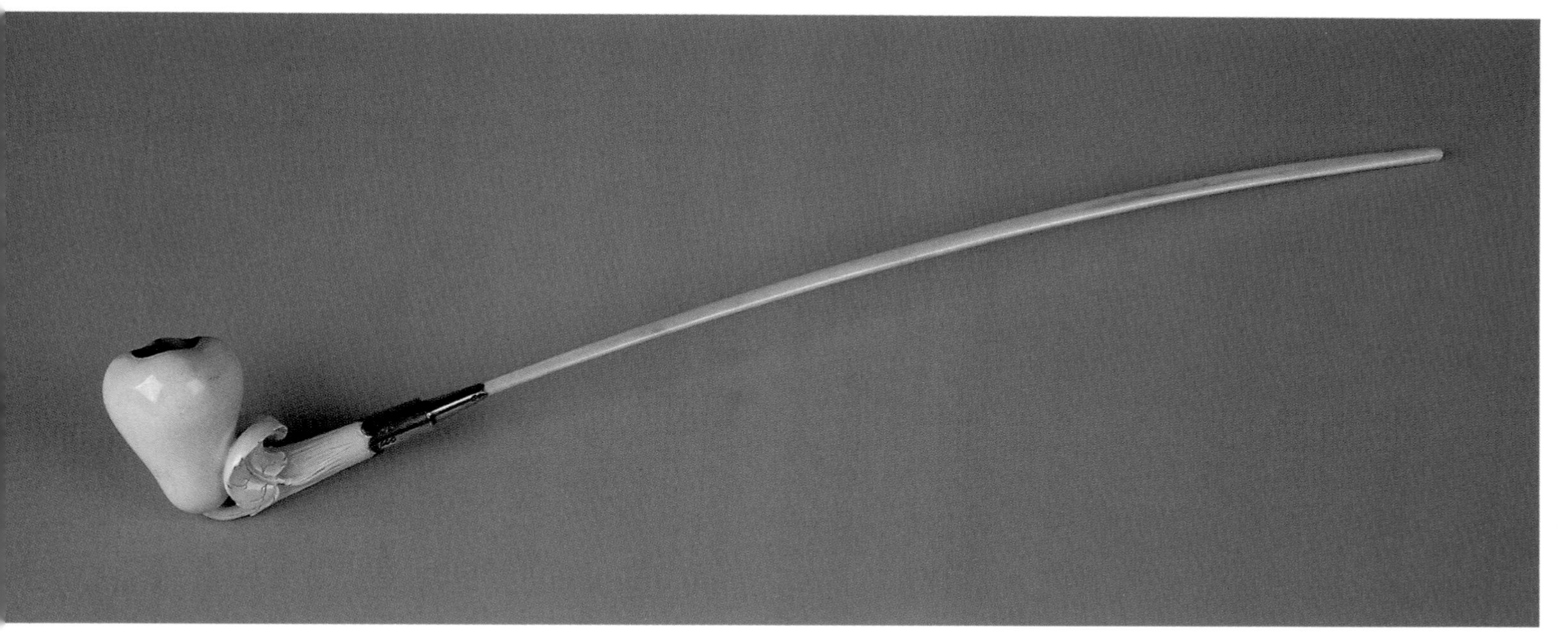

Pipe, flower bud, albatross bone stem, 17.5" l., 2.5" h., French, ca. 1910. *Courtesy of a WC Collector.*

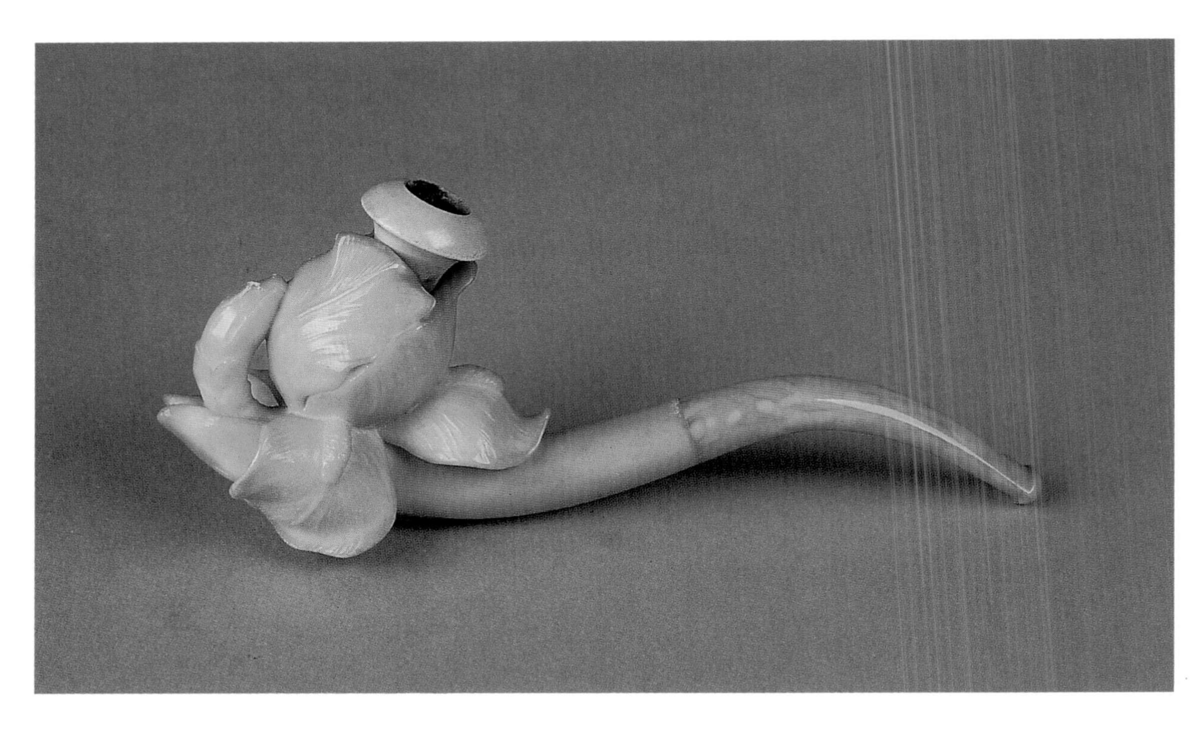

Cheroot holder, flower bud unfurling its petals, 6.5" l., 3" h., J. Sommer, Paris, ca. 1900. *Courtesy of a WC Collector.*

Pipe, figurative head of feral monkey sporting beret, inset glass eyes, jaw distended, mixed media shank of chased silver mounts, meerschaum, and horn mouthpiece, 12.5" l., 2.75" h., prob. French, 1880. *Courtesy of the FB Collection.*

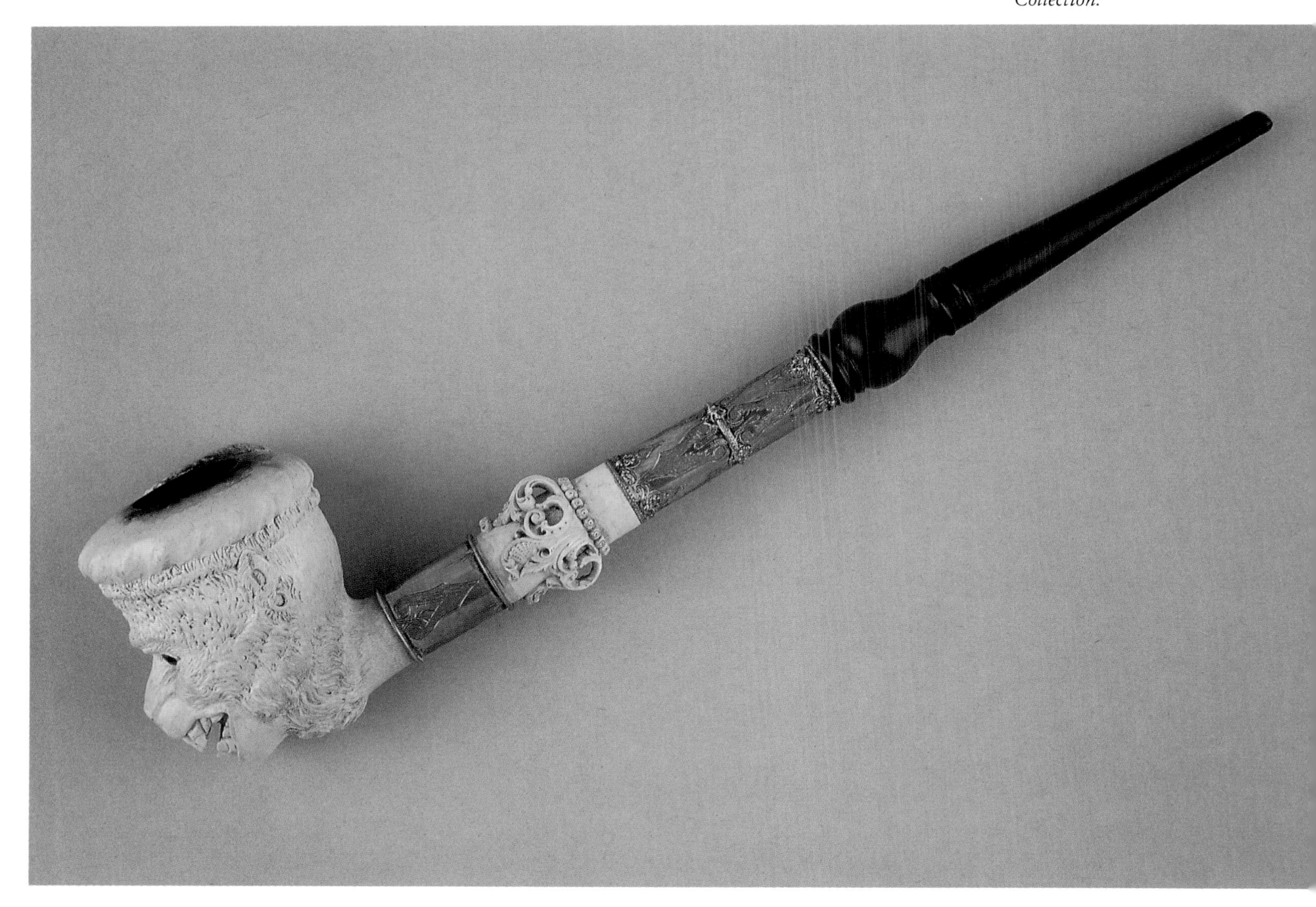

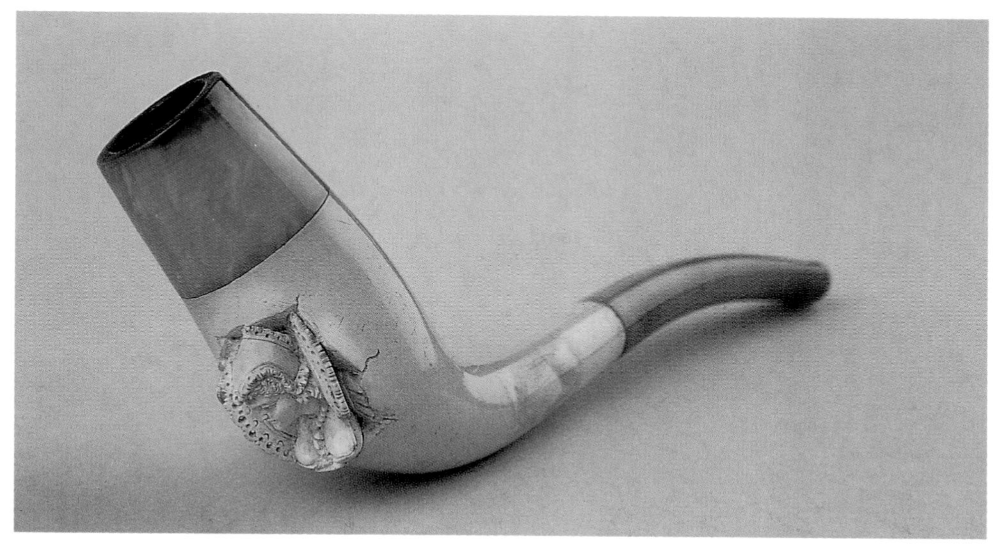

Pipe, clever accent of girl's diminutive head emanating from bowl, amber inset rim, 5.5" l., 3" h., prob. French, ca. 1910. *Courtesy of the Author's Collection.*

Cheroot holder, anatomically correct skeleton holding scythe, after the Grim Reaper, 8" l., 4" h., prob. Austrian-Hungarian, ca. 1880. *Courtesy of the SP Collection.*

Cheroot holder, two chimney-sweeps and washerwoman, 7.75" l., 4.75" h., Emanuel Czapek, Prag, ca. 1880. *Courtesy of the SP Collection.*

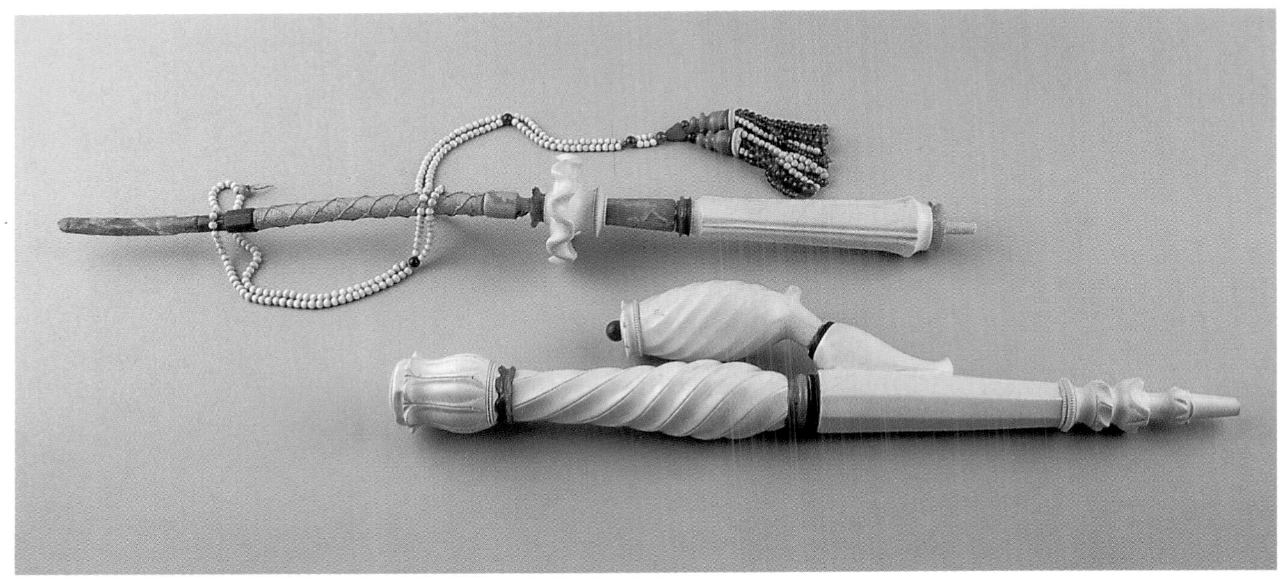

Pipe, *gesteckpfeife*-style, made either for exhibition or as advertisement, hand-turned components in meerschaum, amber accents, flexible hose and amber mouthpiece, 47" h., prob. German, ca. 1900. *Courtesy of the SP Collection.*

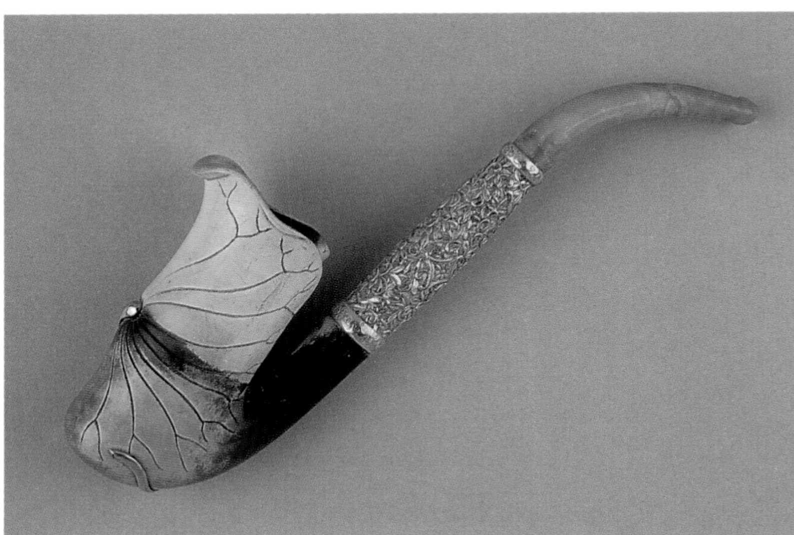

Left:
Pipe, art nouveau, conventional ornamentation, 14-kt rose-gold wire foliate lines incused on bowl, 14-kt yellow-gold floral embossed shank band, amber mouthpiece, 7.5" l. Shank bears date "1883," case is stamped "Tiffany." *Courtesy of the FB Collection.*

Below:
Pipe, Medieval bard, in amber, seated on throne chair, playing lute, 11" l., 5.5" h., C. C., Paris, ca. 1910. *Courtesy of the Author's Collection.*

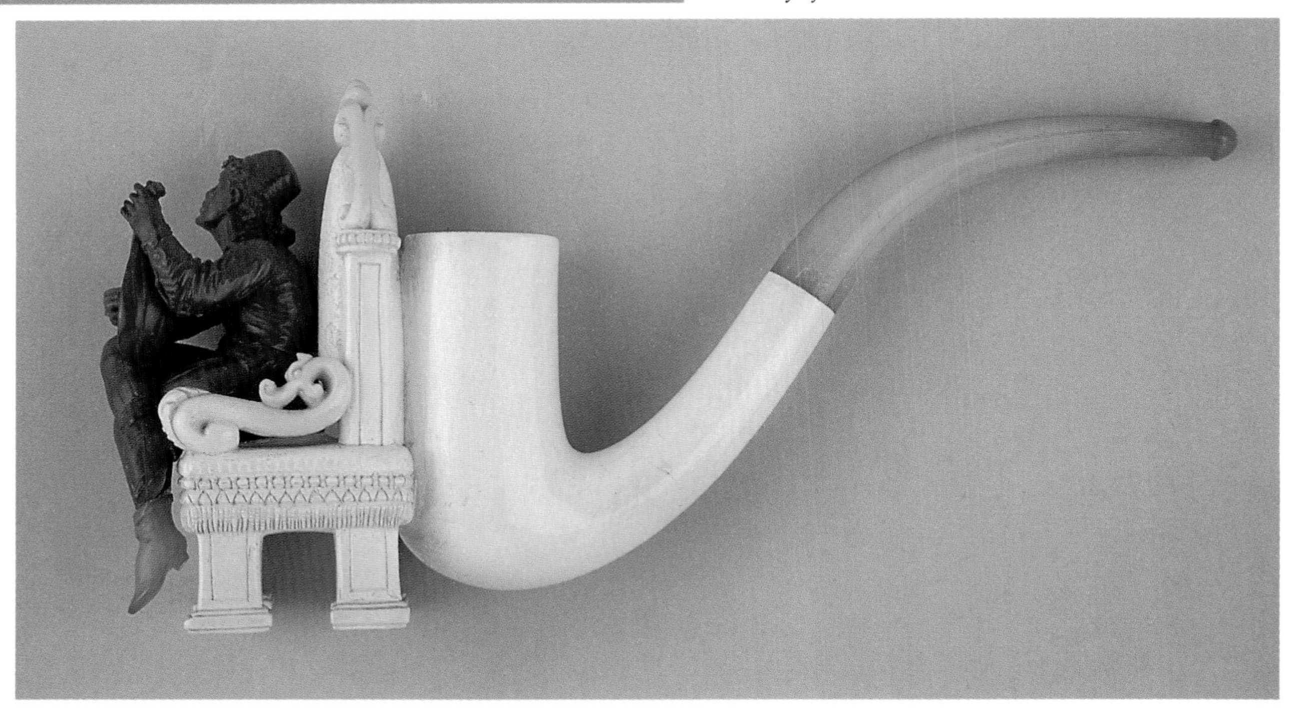

Cigarette holder, whimsical, minstrel seated on amber clef playing banjo, 8" l., 3" h., French, ca. 1910. *Courtesy of a WC Collector.*

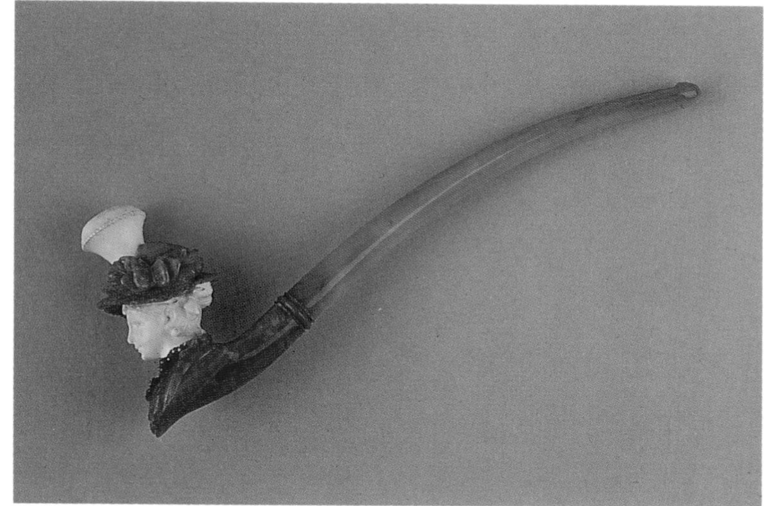

Cheroot holder, figurative head of woman encased in amber, 7" l., German or French, ca. 1900. *Courtesy of the SP Collection.*

Pipe bowl, pre-colored, art nouveau-style, cylindrical shape, amber finial, 4" diameter, prob. Austrian-Hungarian, ca. 1875. *Courtesy of the Author's Collection.*

Pipe bowl, table model, cylindrical shape, band of incised scrollwork circumscribing center, Near East metal filigree motif of lotus-and-palmetto motif, onion dome, 3.5" diameter, prob. French, ca. 1910. *Courtesy of a WC Collector.*

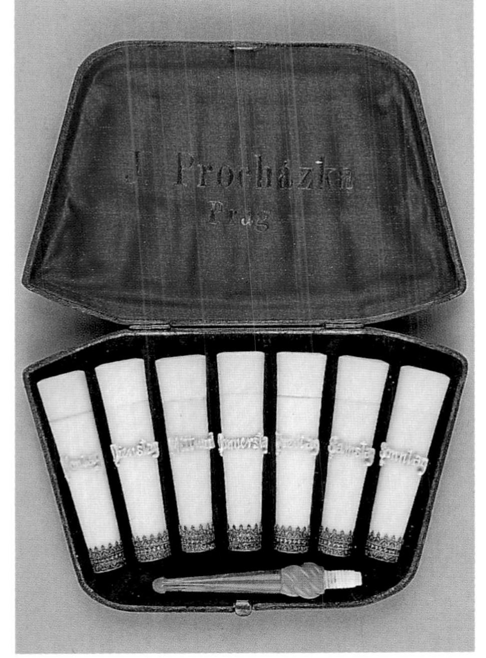

Cigar holder set, cased, 7-day, each day's holder identified in raised (German) block letters, J. Prochazka, Prag, ca. 1900. *Courtesy of the SP Collection.*

Salesman's sample case, assortment of cheroot holders, cigarette holders, and stems, prob. German or Austrian, ca. 1890. *Courtesy of a WC Collector.*

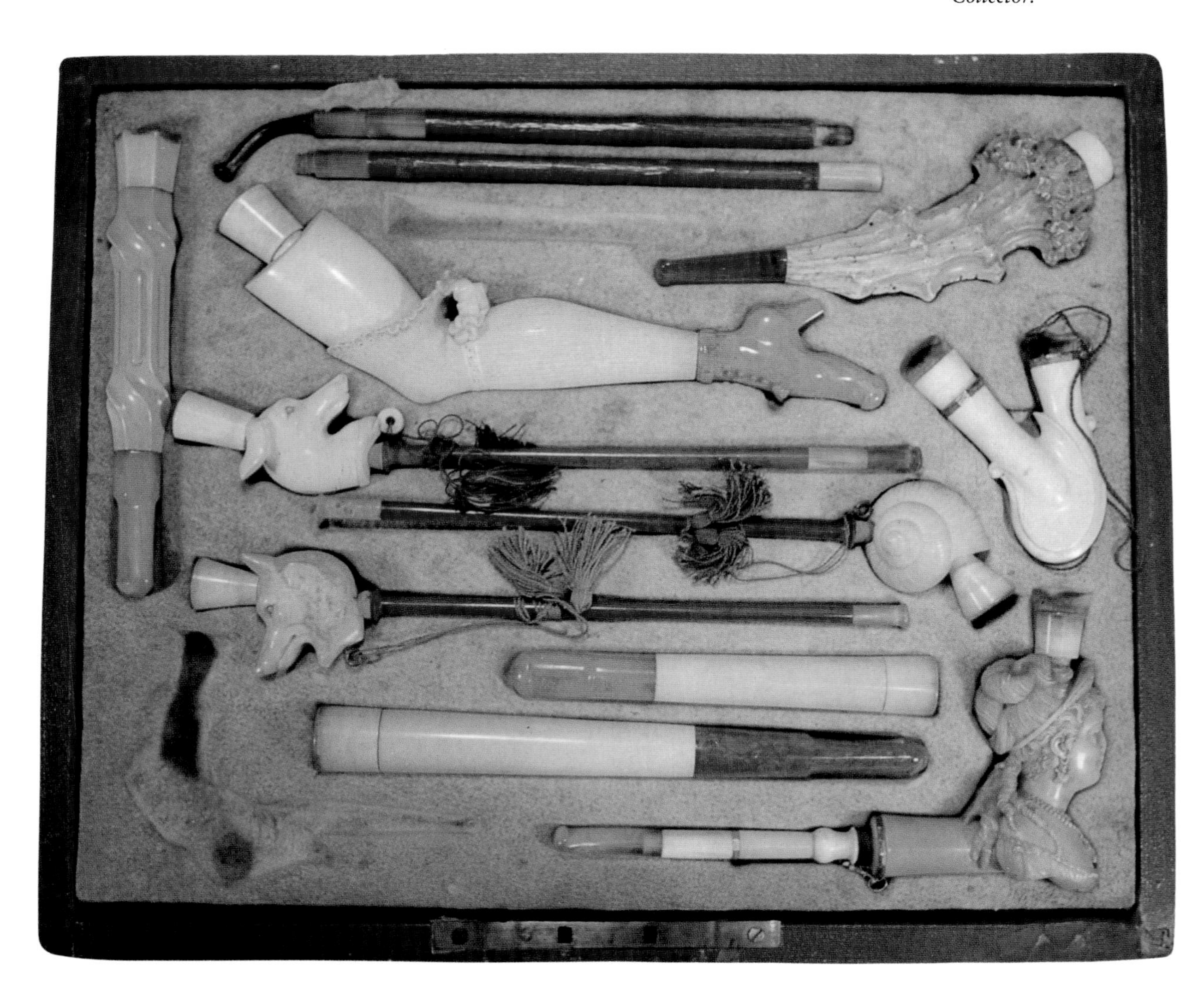

Chapter XV. Naïve, Suggestive, and Erotic

I save for the last chapter those meerschaums that are symbolically naïve, suggestive, and erotic. Every collector has a different interpretation of each of these, so let me offer mine. The naïve represents the unsophisticated, the inexperienced, the awkward, and perhaps the most innocent of amatory motifs. The suggestive portrays subject matter that is evocative and a bit risqué. The erotic arouses strong sexual desire, it is uninhibited allusion. Is this range of motifs commonplace? Yes! These three and the pornographic have been treated in pipes of wood, porcelain, clay, and even metal, so why not meerschaum? Although pornographics seem to command the highest prices, unfortunately, as was mentioned earlier, all those that I have encountered lack fine craftsmanship, their execution is almost always crude and unfinished, and the motifs are limited to one or two variants of sexual coupling. I suspect that, like X-rated movies, pornographic meerschaums might have been "under the counter" sales, special orders not within the standard product line, and executed hurriedly by the less-skilled in the workshop. Only one pornographic cheroot holder has been included as an exemplar. In principle, pornographics are absent from this book for reasons of good taste.

Cheroot holder, prurient motif, recumbent nude flirting with cockatoo, 6.5" l., 2.5" h., Ostpreussische Bernstein Industrie, Berlin, ca. 1890. *Courtesy of the Author's Collection.*

In my opinion, the other three categories of amatory subject matter have a component of subtlety and innuendo that is absent in pornographics. Additionally, the subject matter is more varied, as seen in this small assortment of illustrations. These are not every collector's cup of tea, but, on inspection, some have often prompted extended and animated conversation, produced an occasional chuckle, or raised an eyebrow.

Fundamentally, this genre is composed of all the elusive stuff that drives a collector to continue his quest, because we do not know—and we may never know—the entire gamut and breadth of what this small community of technician-artists, situated in various carving centers in the western world, may have produced in 75 prolific years. This is not a disadvantage, but a distinct advantage, because it offers all collectors the hope that tomorrow, next week, next month, or next year, we will find meerschaums exhibiting new and different motifs, and bigger, more refined, and more exotic examples of the meerschaum carver's sculptured art.

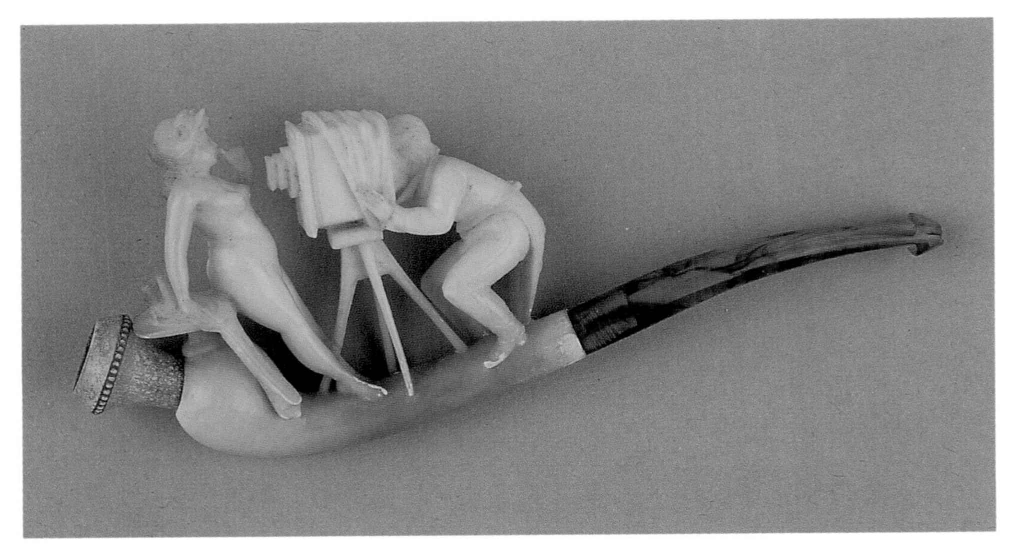

Cheroot holder, photographer capturing nude on celluloid, 6.5" l., 3" h., J.G. Gärtner, Dresden, ca. 1900. *Courtesy of the AZ Collection.*

Cheroot holder, narcissistic woman admiring herself in mirror, hinged apron lifts up to reveal a part of her anatomy, 5" l., 3.5" h., Ferdinand Schnell, Bremen, ca. 1890. *Courtesy of the SP Collection.*

Pipe, narcissistic nude admiring herself in hand mirror, 9" l., 4" h., prob. American, ca. 1885. *Courtesy of the Author's Collection.*

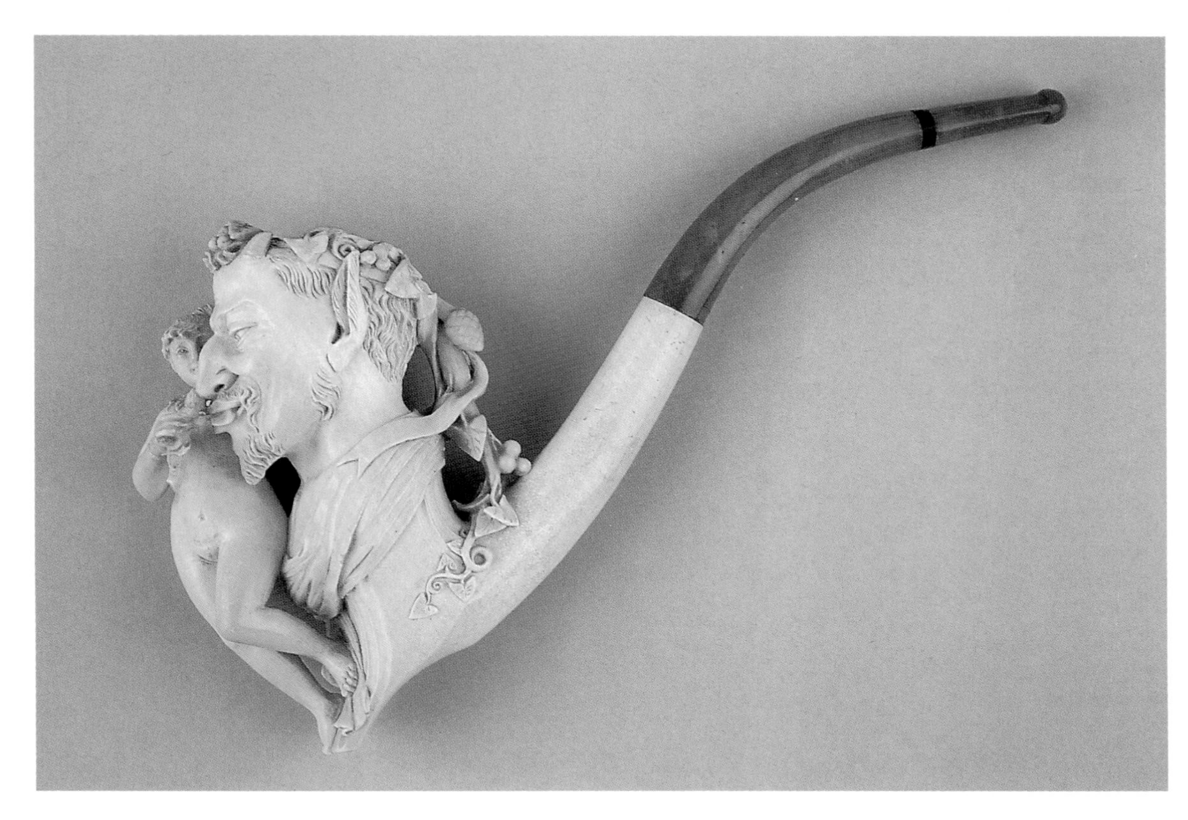

Pipe, figurative head of Satyr, the woodland deity known for rowdiness and lasciviousness, being tempted by nude woman, 10.5" l., 5.5" h., prob. French or Austrian, ca. 1885. *Courtesy of the SP Collection.*

Pipe, frontal bust of turbaned woman revealing her breasts, 6" l., 3.5" h., prob. French, ca. 1885. *Courtesy of the SP Collection.*

Pipe, courtesan reclining in inviting position, attendant cupid with bow, birds on underside, 8" l., 4" h., Bartolomeo, Venezia, ca. 1885. *Courtesy of the SP Collection.*

Pipe, two women flirting with man emerging from under table, 6.75" l., 2.5" h., prob. English, ca. 1885. *Courtesy of the AZ Collection.*

Cheroot holder, two ladies of the night, 10.5" l., 5" h., Franz Hiess, Wien, ca. 1885. *Courtesy of the Author's Collection.*

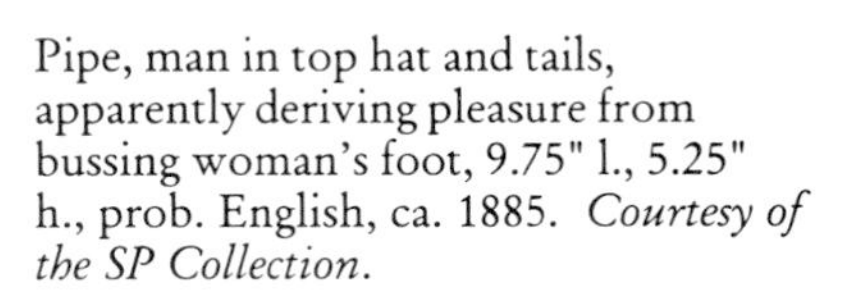

Pipe, man in top hat and tails, apparently deriving pleasure from bussing woman's foot, 9.75" l., 5.25" h., prob. English, ca. 1885. *Courtesy of the SP Collection.*

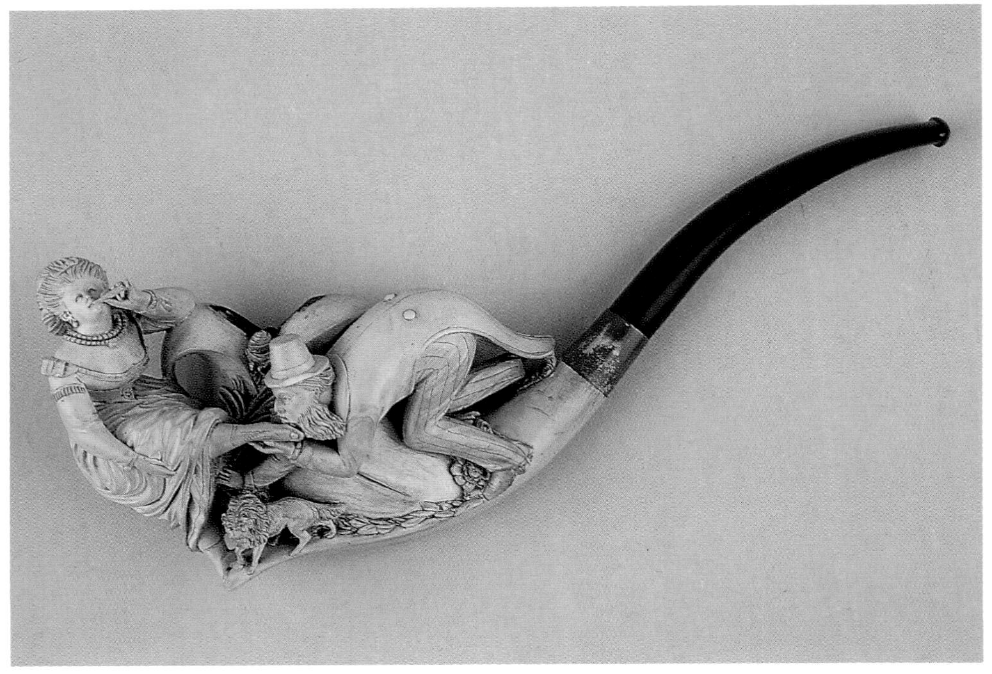

Pipe, young lovers under shelter of fronds being spied by elderly bearded voyeur, 14.5" l., 6.5" h., prob. Austrian or German, ca. 1885. (Perhaps after *Under the Leaf*, a bronze by Albert-Ernest Carrier-Belleuse, ca. 1880.) *Courtesy of the SP Collection.*

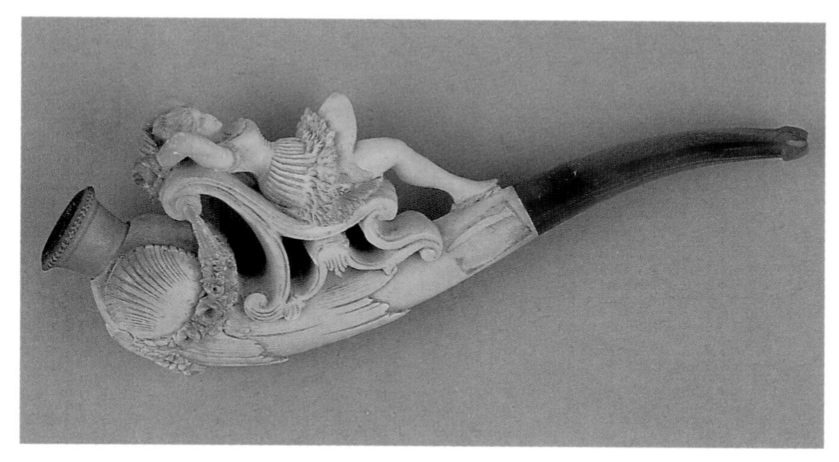

Cheroot holder, woman reclining in very suggestive and inviting pose, 5.5" l., 2" h., Ostpreussische Bernstein Industrie, Berlin, ca. 1885. *Courtesy of the Author's Collection.*

Cheroot holder, nude serenading salacious Man in the Moon, 8.25" l., 3.75" h., Emanuel Czapek, Prag, ca. 1875. *Courtesy of the AZ Collection.*

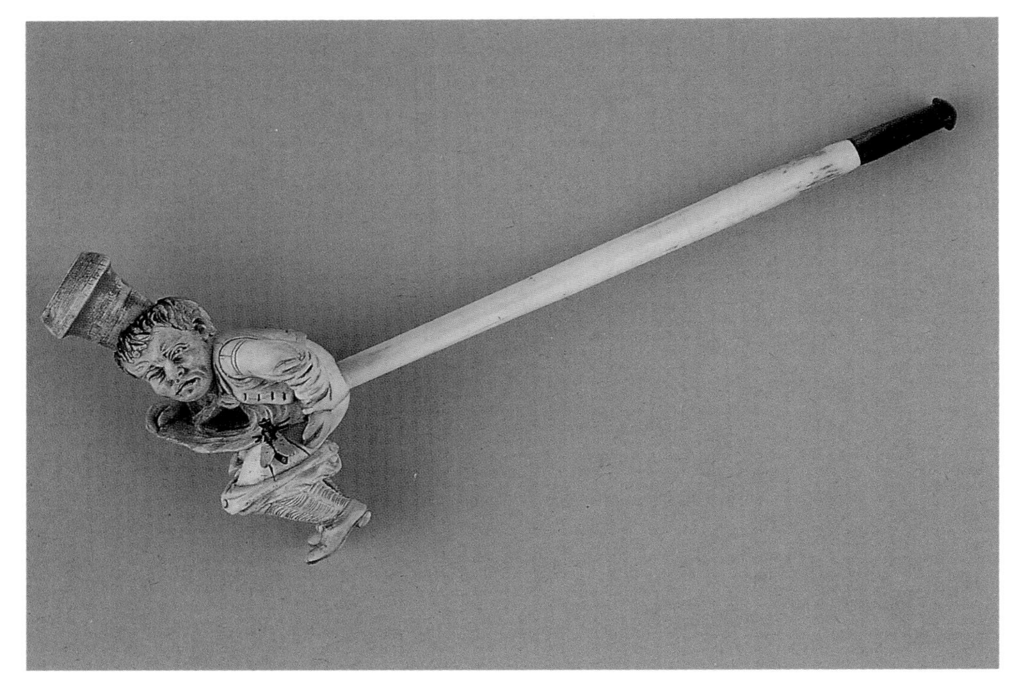

Cheroot holder, young lad, his pants having fallen, shoos away fly, ivory push stem, 6" l., 2.5" h., prob. German, ca. 1900. *Courtesy of the SP Collection.*

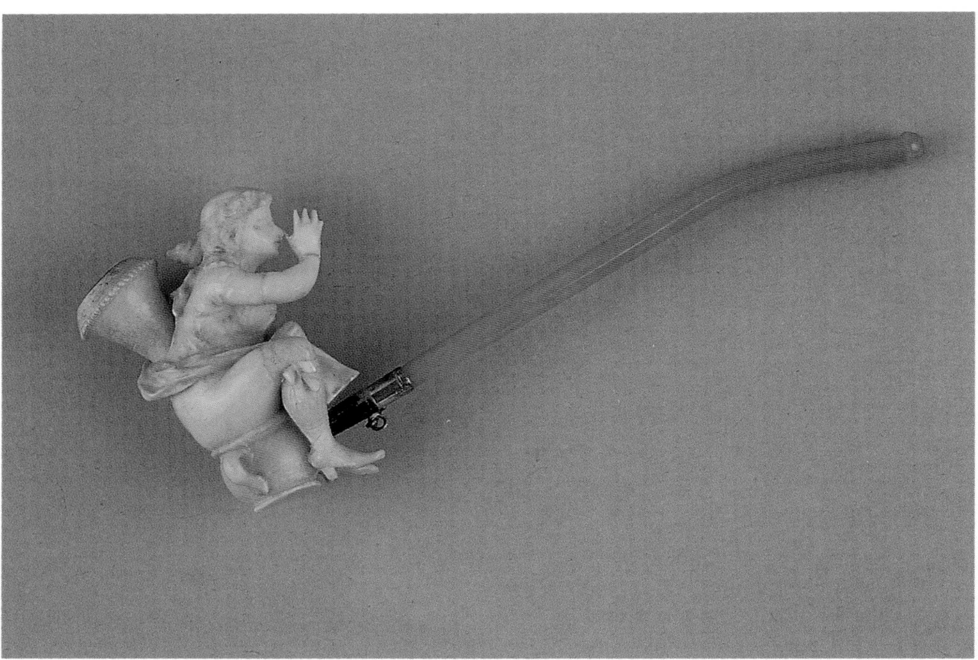

Cheroot holder, woman, seated on chamber pot, thumbs her nose, amber push stem, 5.5" l., 2.5" h., prob. German, ca. 1900. *Courtesy of the SP Collection.*

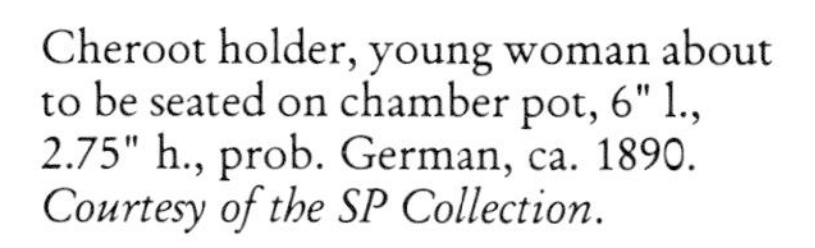

Cheroot holder, young woman about to be seated on chamber pot, 6" l., 2.75" h., prob. German, ca. 1890. *Courtesy of the SP Collection.*

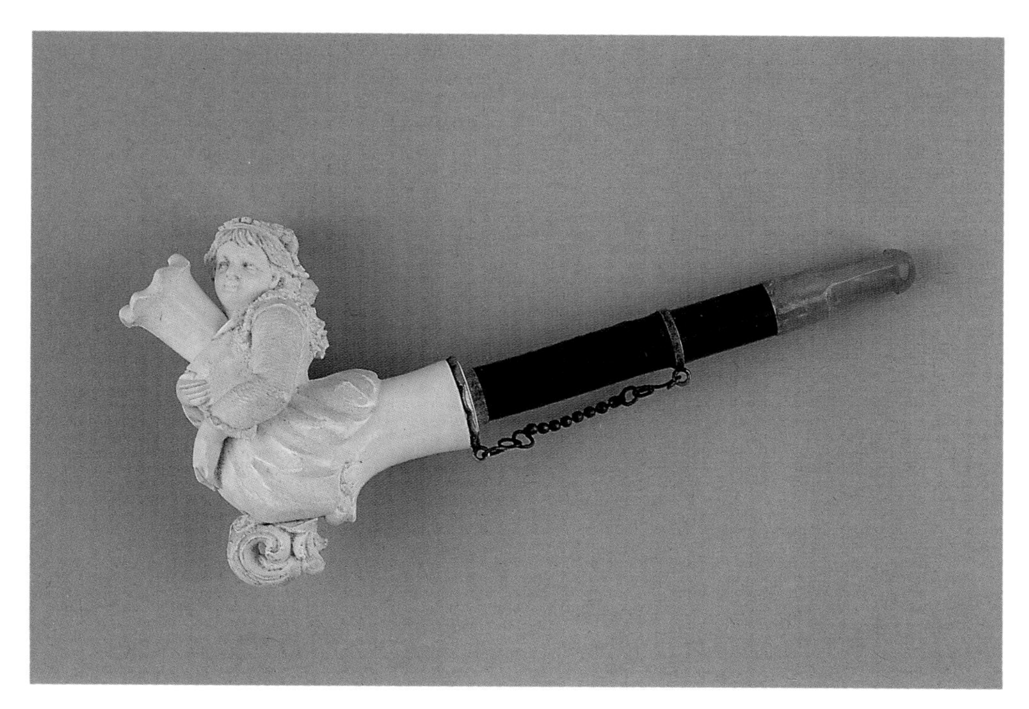

Cheroot holder, bust of woman, cherrywood push stem, removable slide on underside reveals part of her anatomy and indecent phrase inscribed in German, 5.75" l., 2.75" h., ca. 1900. *Courtesy of the SP Collection.*

Cheroot holder, obverse side of woman in snugly wrapped clothing, 2" l., 5" h., ivory push stem, prob. German or French, ca. 1900. *Courtesy of the MG Collection.*

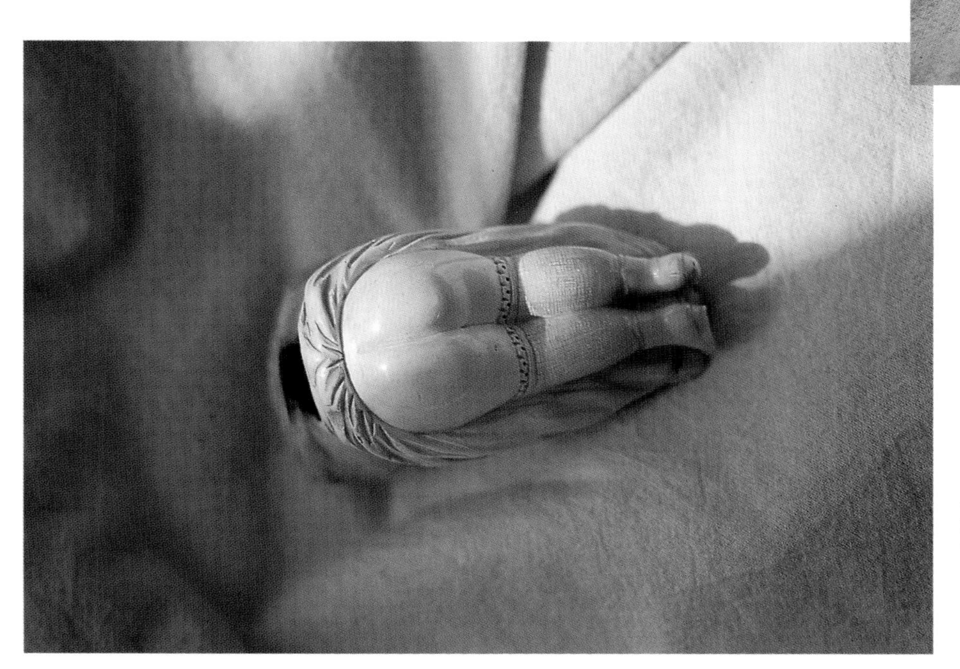

Reverse side of the same cheroot holder.

Cheroot holder set, milkmaid, 2" l., 1.75" h., and monk, 1.75" l., 2.75" h., each configured with hinge at the waist, prob. German or French, ca. 1900. *Courtesy of the JTB Collection.*

With the hinges raised!

Cheroot holder, man and woman in amorous interlude, 5.5" l., 2.5" h., prob. German or French, ca. 1900. *Courtesy of the AZ Collection.*

Endnotes

1. *A Complete Guide to Collecting Antique Pipes*, 1979.
2. When used alone in this book, the word meerschaum is a collective term for antique pipes, cheroot holders, and cigarette holders.
3. "Meerschaum Pipes: How They Are Made, and the Science of Coloring Them Properly," *The New York Times*, July 27, 1874.
4. Leolinus Siluriensis (pseud.), *The Anatomy of Tobacco: Or Smoking Methodised, Divided, and Considered After a New Fashion*, 1884, 17.
5. Cope Brothers & Company, *Pipes and Meerschaum. Part The Third. English and European Pipes*, Cope's Smoke Room Booklet Number Twelve, 1896, 43.
6. Ernst Voges (comp. and ed.), *Tobacco Encyclopedia*, 1984, 200.
7. Meerschaum "may be a German corruption of the word myrsén, which is the native name for this substance in its Asiatic home" (An Old Smoker, *Tobacco Talk*, 1894, 29). In other European languages, meerschaum is écume de mer (French), espuma de mar (Spanish), and schiuma di mare (Italian).
8. Herman Melville, *Mardi: and A Voyage Thither*, 1849.
9. George Zorn & Co., *Pipes, Matches, Walking Canes, and Playing Cards, Illustrated Catalogue*, 3rd edition, c. 1886-87, 23.
10. George Zorn & Co., *Pipes & Smokers Articles, Illustrated Catalogue*, 5th edition, c. 1892. Reprint, 1989, 54.
11. Gardner D. Hiscox, M.E. (ed.), *Henley's Twentieth Century Book of Formulas, Processes and Trade Secrets*, 1957, 469.
12. *Tobacco Talk*, op. cit., 29.
13. Carl Avery Warner, *Tobaccoland*, 1922, 415-416.
14. George Zorn & Co., *Illustrated Catalogue*, 3rd edition, op. cit., 23.
15. Constantin, *Biographisches Lexikon des Kaisertums Österreich*, 1874, 396. The English translation appears in Ferenc Levárdy, *Our Pipe-Smoking Forebears*, 1994, 119.
16. Robert Cudell, *Das Buch vom Tabak*, 1927, 144-147. The English translation appears in *Our Pipe-Smoking Forebears*, op. cit., 118-119.
17. *Our Pipe-Smoking Forebears*, ibid., 119.
18. See References for complete bibliographic data.
19. E. Reid Duncan, "Mysteries of Meerschaum," *Pipe Lovers*, Vol. IV, No. 1, January 1949, 8.
20. Idem.
21. Even if this story is a complete hoax, the pipe is real, now the property of a Chicago-area collector, and a photograph of the Brigand is contained in this book.
22. "Meerschaum," *Cope's Tobacco Plant*, No. 59, Vol. I, February 1875, 706.
23. The eight titles, Bock, Koppa, Meyer, Raufer, Reineke, Tomas, Tomasek, and Ziegler, are listed in References.
24. "Meerschaum II," *Cope's Tobacco Plant*, No. 60, Vol. I, March 1875, 719-720.
25. "Meerschaum IV," *Cope's Tobacco Plant*, No. 63, Vol. I, June 1875, 755.
26. Edward Vincent Heward, *St. Nicotine of the Peace Pipe*, 1909, 197.

27. Bevis Hillier, "Pipe Dreams," *Punch*, August 1983, 184. Hillier (and Heward) were not alone in the belief that Vienna was the center of the meerschaum industry from about 1860 to 1900, but this observation is not completely supported by the table in Annex A. Perhaps, just as the story of the Hungarian cobbler might be a trifle exaggerated, Vienna may have been the singular hub of the trade for a brief period in the 19th century, and much later it became just another meerschaum carving center.
28. W. A. Penn, *The Soverane Herbe. A History of Tobacco*, 1901, 167.
29. G. M. Raufer, *Die Meershaum- und Bernsteinwaaren-Fabrikation. Mit einem Anhange über die Erzeugung hölzerner Pfeifenköpfe*, 1876, 4. Translated literally, Vienna is and will remain the central point of manufacture; Vienna was the cradle from which meerschaum emerged as an article of fashion.
30. Fritz Morris, "The Making of Meershaums," *Technical World*, April 1908, 194-195.
31. The exposition opened in the spring of 1901 and attracted nearly 8 million people. Sadly, on September 6, the event was marred when an anarchist, Leon Czolgosz, shot President McKinley at point-blank range outside the Temple of Music on the exposition grounds, and the President died one week later.
32. "Buffalo Pipe Carver For Half Century Upholds Traditions of Famous Family," *The Buffalo Times*, March 24, 1934.
33. If you look very closely at the photo on the bottom of page 30, the Battle of Bunker Hill pipe is ensconced in the storefront window. In its day, the pipe was claimed to be the largest and most exquisitely carved meerschaum in the world and, according to a brief clipping in *The Boston Herald*, December 21, 1906, was worth $40,000.
34. Museum of Tobacco Art and History, *Meerschaum Masterpieces. The Premiere Art of Pipes*, December 1990, 3.
35. Amoret Scott, and Christopher Scott, *Tobacco and the Collector*, 1966, 34.
36. B. Linger Company of Vienna, Austria and London, *Illustrated Catalogue. Wholesale & Export Only*,ca. 1900, n.p.
37. *Our Pipe-Smoking Forebears*, op. cit., 129.
38. According to its ca.1899 catalog, Salmon & Gluckstein Ltd. claimed to be the largest European retail tobacconist of the late 19th century, with head offices in London and more than 120 retail branches throughout the United Kingdom. The company imported its meerschaum pipes and cigar tubes (known as cheroot holders in the United States) from various carving centers in Europe and merchandised them in fitted cases stamped with the S&G logo inside.
39. Frederick W. Fairholt, F.S.A., *Tobacco: Its History and Associations: Including an Account of the Plant and Its Manufacture; With Its Modes of Use in All Ages and Countries*, 1859, 222-223.
40. Brian Innes (comp.), *The Book of the Havana Cigar*, 1983, 98-99.
41. H. F. & Ph. H. Reemtsma, *Tabago. A Picture Book of Tobacco and the Pleasures of Smoking*, 1960, 81.
42. J. W. Cundall, *Pipes and Tobacco Being A Discourse on Smoking and Smokers*, 1901, 33.
43. Roger Fresco-Corbu, "Mystic Meerschaum," *Art & Antiques Weekly*, Vol. 9, No. 2, January 20, 1973, 27.
44. Anon., *An Arm-Chair in the Smoking-Room: Or Fiction, Anecdote, Humour, and Fancy For Dreamy Half-Hours. With Notes on Cigars, Meerschaums, and Smoking, From Various Pens*, 1868, 342.

45. George Zorn & Co., *Illustrated Catalogue*, 3rd edition, op. cit., 29.
46. *Meerschaum Masterpieces*, op. cit., 5.
47. John E. Fielding (ed.), *Historical Dictionary of World's Fairs and Expositions, 1851-1988*, 1990, 376-377.
48. U.S. Centennial Commission, *International Exhibition, 1876.* Official Catalogue, 1876, viii, 105.
49. Almost 130 years later, this pipe, once the property of Astley's, London, is now owned by a West Coast collector.
50. "Exhibition of Pipes at South Kensington," *Cope's Tobacco Plant*, No. 18, Vol. I, September 1871, 212.
51. "Meerschaum III," *Cope's Tobacco Plant*, No. 61, Vol.I, April 1875, 780.
52. *Tobacco: Its History and Associations*, op. cit., 196.
53. "Tobacco-Pipes: With Special Reference to the Collection in the London International Exhibition, 1873," *Cope's Tobacco Plant*, No. 44, Vol. I, November 1873, 528.
54. "Meerschaum IV," op. cit., 754.
55. Idem.
56. *The New York Herald*, January 31, 1867.
57. *International Exhibition, 1876*, op. cit., 128.
58. *Treasures of Art, Industry and Manufacture Represented in the American Centennial Exhibition at Philadelphia*, 1876.
59. *The New York Herald*, op. cit. Prices, on average, for median-quality meerschaum pipes in New York City at that time ranged from a low of $1.50 to as much as $500. The article points out that cigar holders also commanded high prices, often as much as $74, depending on the style of carving. These prices, by today's standards, were high.
60. "Meerschaum and Amber," *The Scientific American*, XV:II, September 8, 1866, 161.
61. Sarah Wilford, "10,000 Pipes—at Every Price," *Long Island Press*, January 24, 1936, reported that the Columbus pipe, manufactured in the late 1800s, was now valued at $50,000.
62. Findling, op. cit., 395-401.
63. "The Pipes of Various Races," *Cope's Tobacco Plant*, No. 10, Vol. I, January 187, 113.
64. Edinburgh Museum of Science and Art, Science and Art Department, *Guide to the Loan Collection of Objects Connected With the Use of Tobacco and Other Narcotics*, 1880, 6.
65. See Benjamin Rapaport, "Pipes of Our Presidents," *Antiques & Collecting Magazine*, November 1994.
66. Harold Littledale, "Meerschaums," *U.S. Tobacco Review*, Fall 1979, 5.
67. "Mystic Meerschaums," op. cit., 25.
68. The highest price on record for an antique meerschaum pipe was $41,000 at the London auction gallery of Sotheby's in November 1990. The pipe depicted 'Androcles and the Lion,' the legendary Roman slave who was spared in the arena by a lion from whose foot he had long before extracted a thorn.
69. *Tobacco and the Collector*, op. cit., 34.
70. "Pipe Dreams," op. cit., 184.
71. Meerschaums with pornographic motifs are found in these two books: Cotti Burland, *Erotic Antiques or Love is An Antic Thing* (Lyle Publications, Selkirkshire, Scotland, 1974), and Jean-Pierre Bourgeron, *Les Masques d'Eros. Les Objets Erotiques de Collection à Systeme* (Les Editions de l'Amateur, Paris, France, 1985).

ANNEX A. Makers and Marks

This table is a compilation of names associated with meerschaum and amber manufacture in the 19th century to the early 20th century. All known data, including the company's various recorded name changes, are included. I am neither confident that the spelling is correct in all instances, nor am I sanguine that all the names represent carvers; I am certain, however, that this list is not complete. Often, a retailer, wholesaler, distributor, or importer/exporter requested that his name be embossed in the case rather than the name of the carver (or he superimposed a paper label of his name over that of the carver). Hence, the attribution of a particular pipe to a specific carver is not ever certain or assured. Note the distribution of names on both sides of the Atlantic; during this period, the majority of these are, not surprisingly, European.

The contents represent many years of accumulated notes gleaned from pipe and cheroot cases and some support from S. Paul Jung of Bel Air, Maryland, a leading researcher of the American pipe industry and a collector of clay tobacco pipes. He has created a "living" database of all U.S. pipe-makers of this same period by researching both business and city directories which will be published as *Tobacco Pipe Manufacturers and Distributors Found in U.S. Directories in the Library of Congress* sometime in 1999. He graciously allowed me to compare my notes with his, and some of his data about American meerschaum pipe companies appear herein. Paul took no license or liberties in his research; if the directory listed "smokers' articles" or "pipes," he extracted this exactly without embellishing the entry with words such as briar or meerschaum. But, if the words "meerschaum," "meerschaum pipes," or "meerschaum pipe manufacturer" appeared in the business directory, I included that name or company in this table. It is likely, then, that had a more precise or elaborate description of every pipe manufacturer or retail tobacconist of the period operating in the United States appeared in these directories, this list might have been yet longer.

COUNTRY	CITY	NAME(S)
Austrian Empire/ Kingdom of Hungary/ Austria-Hungary	Bratislava	Csang Hermann Rappel Schulz
	Debrecen	Dániel Schwartz
	Eger	Hoffman
	Innsbruck	T. Lorenz
	Marienbad (Marianske Lazne)	Simenon
	Pest/Budapest	Philip Adler and Sons (Adler Fülöpes Fia) & József Adler u. Medetz Dávid Albachar Ignác Berger Bernát Bleyer Sándor Dávid Donath Feldsberg Gallwitz János Gasparek Sándor Keveházi; Zoltán Keveházi F. & Henrik Martiny József Medetz Dávid Minkus Nemetz Novágh Oppenheimer Sebestyén Pésci Rettich Roth Richárd Schmidt Adolf Weck S. Weiss (Weisz), es Fia Ármin Weisz Sándor Ziegler Ignác Zsigmond
	Pilsen (Plzen)	Raim Jindra
	Prag/Praha	Bruder Czapek; Emanuel u. M. Czapek Jos. Dolezal J. Procházka
	Szombathely	Bruckner
	Trencen	Joseph Rettinger
	Wien	Androsch & Eichler Leopold Auteried

COUNTRY	CITY	NAME(S)
	Wien (cont.)	B. Backofen Bauer & Pokorny A. Berger Johann Brachtl Sigmind Brill Johann Brix Franz Czerwak Johann Dauda A. deLambert Wtw. Diessl's & Müller Stefan Drenzel Johann Edhofer Heinrich Eyer u. Söhne Joseph Fischer Hermann Flöge Johann Friedrich Mathias Fuchs Peter Johann Nepomuk Geiger Moritz Goldmann Josef Gordon Gorlitzer Brüder Gross Gumperdorfer Strasse Werke Bernhard Guth Ludwig Hartmann & Eidam (K.K. Privilegirten Drechslerwaaren-Fabrik) Franz Hiess; Franz Hiess & Söhne Karl (Károly) Hiess Bernhard Hirschler Joachim Hitschmann Rudolf Hoffmann M. Hrauda Julius Hübsch Friedrich Jacobi u. Söhne Gebrüder Kanitz F.R. Kast Peter Keiss Joh. Nep. Kirchner Albin Klitsch August Kloger Moritz Knopler Carl Kober Koch & Co. Georg Koppa Josef Koritzer Johann Korotin J. Kosira Emanuel Köstner Josef Krammer Jos. Lang & Co. Johann Lederer B. Linger

COUNTRY	CITY	NAME(S)
	Wien (cont.)	Adolf Litchblau; Jakob Litchblau; Rudolf Litchblau
		A Lüthe
		H. Mager
		August Maier
		Josef Mailer
		A. Marusch
		G. C. Matthess
		Johann N. Menhard
		C.W. Möller, Fürstlicher Hoflieferant; Gebrüder C.W. Möller
		Johann Müller
		Leopold Müllner
		Leopold Nagl
		Simon Nolze
		M.A. Nowatny
		Raimund Raunegger
		Ignatz Reichenfeld
		P. Reiss
		G. Roso
		Eduard Rothe
		L. Rothenstein
		Brüder Rothmiller
		Joseph Ruger
		Heinrich Schilling
		Alois Karl Schmidt
		Isador Schmitz
		Ed. Schubert, Vorm. Nolze
		Eduard Sievert
		Anton Strobel
		Josef Strelitz
		Arnold Trebitsch
		J. Vegiato
		Franz Weber
		C. Weis
		Weiss & Rödel
		Johann Weissenberger
		Jac. Werner
		Fr. M. Wolff/r/M.L./Carl Beit
	Unknown	Motto Bilsenkraut und Mohn
		Deáky (Dëaki or Tiagi)
		Julius Kuszmann
		Emil u. Etelka Spiro
Belgium	Anvers	Hoffmann, Fabricant de Pipes
		Lievin Steppe, Fabrique de Pipes
	Bruxelles	E.R. Rabe
		A^{te} Vernaillen
	Bruxelles (cont.)	C.B. Vinche; Victor Vinche; J. B. Vinche, Fabrique de Pipes "Au Nabab"
		D. Vroedsie
	Liège	Massin-Bage, Fab^{t}
Ceylon	Colombo	D. Macropolo
Denmark	København	Wilh. Jørgensen
		Carla Nielsen
England	Birminghan	W.H. Newman Ltd., Pipe Makers
		J. Underhill
	Cambridge	A. Colin Lunn
	Leeds	Singleton & Cole, Ltd.
	Liverpool	George Watts
	London	B.R. Arkell
		William Astley & Company, Ltd.
		BBB
		D. Brumfit (Piccadilly) Ltd.
		Charles Edwards; F.E. & C.; F. Edwards Manufacturers; F. Edwards & Co., Mfgrs.
		G. Fauerbach
		Adolph Frankau & Company; A.F.C.
		Fribourg & Treyer
		Joh. Friedrich
		Charles Goas & Co.
		A. E. Kettle
		J. Koppenhagen; Koppenhagen Bros.
		B. Linger
		Litsica, Marx & Co.
		H. Perkins
		Salmon & Gluckstein
		S. Weingott & Sons
	Manchester	R.J. Lea
	Newcastle-on-Tyne	Robert Sinclair, Meerschaum & Briar Pipe Mfgr.
France	Bordeaux	Alphonse Lancel
		Manufacture Française d'Armes et Cycles de Saint-Etienne (Loire)
	Colmar	Charles Harnisch

COUNTRY	CITY	NAME(S)
	Gand	Ph. Caron
	Havre	Fabrique de Pipes A. Benoit
	Le Mans	Au Pacha
	Lille	Au Phénix Manufacture Française d'Armes et Cycles de Saint-Étienne (Loire)
	Lyon	Au Turc, Goob Manufacture Française d'Armes et Cycles de Saint-Étienne (Loire)
	Marseille	Alphonse Lancel Au Phénix
	Nancy	Manufacture Française d'Armes et Cycles de Saint-Étienne (Loire)
	Nantes	Manufacture Française d'Armes et Cycles de Saint-Étienne (Loire)
	Nice	Kandler, Sous les Arcades
	Paris	J.F.A. A.B., Fabrication Parisienne, Écume et Ambre Sublimes à la Bonne Pipe à la Grande Pipe à la Pipe de Chasse (successeur: M. Drot) à la Pipe de Paris à la Pipe du Nord (Emile Schomogué) à la Sans Rivale à la Vraie Écume (M. Reynier) à l'Algérien, Fque de Pipes Écume & Ambre, Alexandre Martin Fabricant à l'Egyptien à l'Etoile d'Ambre à l'Indien, J. Gay à l'Oriental Au Bloc d'Ambre, Maison Guyot; H. Guyot; Fabrique de Pipes; H.G. Fab. Au Bosphore

COUNTRY	CITY	NAME(S)
	Paris (cont.)	Au Caid (Maison Hiltz) Au Cosmopolite (successeurs: M. Auerbach; M. Arrachart; M. Legros) Au Fumeur, M. Skopec Au Fumeur Suisse (M. Spiro) Au Grand Turc (M. Testard) Au Grand Vizir Au Khedive (M. Glink) Au Nabab (M. Pati) Au Pacha, Fabrique de Pipes d'Écume (successeur: Maison Lenouvel); Au Pacha, Desbois & Weber Succ; Desbois et Cheville (successeur: Maison Lenouvel) Au Petit Pacha Au Phénix Au Shah de Perse, M. Guillaume Korner Au Sphinx Au Succès Au Turc, M. Gemeiner; Au Turc, Goob Aux Mines d'Écume (M. Lejeune) Aux Pipes Ropp Aux Trois Étoiles Aux 100,000 Pipes G. Boese Bondier, Ulbrich & Cie. G. Bonneau (successeur: à l'Innovateur, M. Balagny) G. Brelet E. Cardon Cawley et Henry (successeurs: Wolf & A. Mathiss) A. deLambert Derger Adrien Dubois de Laval (successeur: J. E. Bailly) Emlekul Jules Fex (J.F.A.); E.F.A. (Emile Fex Aîné); Au Scheriff (J.F.A. Fex) Georges Jeune Flavien (successeur: d' Hopfer et Turc) Fléchet J. Foegly M. Fred

COUNTRY	CITY	NAME(S)
	Paris (cont.)	Frizon
		Pierre Fuchez; BF Constantinople; Bernard Fuchez; Fuchez Fréres (successeurs: H.W.B.)
		Ganneval, Bondier, & Donniger (GBD) F^{ts} à Paris (Maréchal Ruchon et Cie)
		Garnier
		M. Henri Gérin; Henri Gérin & J. Foegly
		Louis Goetsch, Fabrique de Pipes en Écume de Mer, G^{de} Cie; (successeur: Hoffman)
		Auguste Goltsche (successeurs: Goltsche, Henriette X, and Louis Lavisse)
		H. Hautbolt, Fab.
		M. Heimann
		Horak
		Kirby-Beard
		Krebs, Fabricant
		LMB
		Lacanaud
		Fabricant L.L.
		Alphonse Lancel
		Landolff, à l'Epoque
		Langlois
		Latzkó
		E. Laumond
		Edouard Launay
		Leclerc 28 (successeur: Pépin de Graef)
		Legros
		H. Lemaire
		Lichtblau (successeurs, S. & E. Blum [S.E.B.])
		M.M. Aleiter de Vienne
		Maison Mathiss
		Manufacture Française d'Armes et Cycles de Saint-Étienne (Loire)
		Maréchal & Bine (successeurs:B.H.V. [Ruel Jeune et Vve Ruel])
		Morin
		Normand & Foulon; Normand et Schroeder; Normand Frères
		Pajot
		A. Pandevant et Roy

COUNTRY	CITY	NAME(S)
	Paris (cont.)	Georges (& Michel) Petit
		Pettice
		Philippart-Moulin
		Pioche et de Delgas (Garnier)
		J. Schomogué et A. Bichet (successeur: Morin; successeur: J.J.B.)
		G. Schneider, Fabrique de Pipes
		D. Silberberg
		Societé Française des Pipes Kock
		Sommer Frères, Aux Carrieres d'Écume; Sommer Fab. Faivret S.; J. Sommer, Fab^{t}
		Arnold Trebitsch
		Verry
		J. Walter
		Weber
		E. Robert-Weill
		Willfort de Vienne
		Wolff
		F. Zieg
		M. Ziegler
	Rennes	Aide
		I. Sexer
	Rouen	Manufacture Française d'Armes et Cycles de Saint-Etienne (Loire)
	Saint-Etienne	Chevalier
	Toulouse	Manufacture Française d'Armes et Cycles de Saint-Etienne (Loire)
	Tours	Fabrique de Pipes, Fourrs
		Manufacture Française d'Armes et Cycles de Saint-Etienne (Loire)
India	Calcutta & Bombay	D. Macropolo
Ireland	Dublin	Kapp Bros.
Italy	Chiavari	G. Montani
	Firenze	Giuseppe Hlawatschek; Sommer e. Hlawatschek, S.A.
		E. Fratelli Parenti
		Zannone

COUNTRY	CITY	NAME(S)
	Genova	F. G. Kunz F. Savinelli Leopoldo Weiss
	Milano	Edoardo Flegel G. Lightenstern Miccio & Co. Mauricio Pisetzky A. Savinelli
	Napoli	A. Fiedler L. Gambarini Miccio & Co.
	Roma	à l'Armenien Neyrat Piccioni & Co. Luigi Tissioti Fratelli Vittori
	Torino	E. Blanchard, Gio Busso Succ. Maurizio Fürst G. Gremo, Suc. G. Busso Ditta Luciano Giacomo Strauss Pietro Vizio
	Trieste	L. Baumgartner G. Weiss
	Venezia	Bartolomeo Maddalena Bianchi Lod co Jona Fratelli Righini Luigi Vogini
Netherlands	Amsterdam	P. Hassoldt J. Riethof Galerij H. Wasmann Jr., Kunstdaajer
Norway	Lillehammer	G. Larsen
Ottoman Empire	Constantinople	M. Saury
Philippines	Manila	Adad & Ticard La Esmeralda
Prussia/Germany	Bad Kissingen	J. Gillis Östpruessische Bernstein Industrie
	Berlin	C. W. Möller, Furstlicher Hoflieferant; Gebrüder C. W. Möller
	Berlin (cont.)	Östpruessische Bernstein Industrie C.R. Parthum F. R. Rosenstiel, Hoflieferant J. Woythaler Bernstein Fabrik
	Bonn	Heinrich Gremme
	Bremen	C. Schnally Ferdinand Schnell
	Breslau	Östpruessische Bernstein Industrie
	Cassel	Christoph Cotthard
	Danzig	J. Woythaler Bernstein Fabrik
	Dresden	J. G. Gärtner, inhaber Oscar Rüger Franz Hauboldt Georg Koppa, Hoflieferant, K.K. Honeit-Kronprinz v. Preussen Gebrüder Muschweck, Meerschaum und Lederwaren Fabrik; Fabrik Gebr. Muschweck Östpruessische Bernstein Industrie Emil Pietzsch Oskar Ruger
	Frankfurt	J. Gillis, Östpreussische Bernstein Industrie Jakob Hellman Johann Wolfgang Wagner
	Freiburg	E. Brack Jun
	Hamburg	Östpruessische Bernstein Industrie Detl. H. Ritter Ferdinand Tesch
	Hannover	A. Haupt Alg Koester Östpruessische Bernstein Industrie Ferd. Vieregge
	Höhr bei Coblenz	J. Schilz-Müllenbach Pfeifen-Fabrik & Export

COUNTRY	CITY	NAME(S)
	Karlsbad	Joh. Korotin Carl Reichel, Meerschaum & Bernsteinwaaren Erzeuger Joseph Ruger A. & C. Schmidt
	Kissingen	A. &. C. Schmidt
	Köln	Rudolf Gotz Sienem L. Dr. Anselm Gotzl Östpruessische Bernstein Industrie
	Leipzig	Arthur Schneider
	Leipzig-Plagwitz	Versand-Geschäft Mey & Edlich, Koenigl. Sächs., Hoflieferanten
	Mannheim	Josef Gordon
	Mulhausen	B. Holzer
	München	A. Diessl, Koenig Hoflieferant Östpruessische Bernstein Industrie J. Zimmerman, Hofdrechsler
	Nürnberg	M. Held
	Osnabruck	J.C. Zangenberg
	Pappenheim	Michael Strassner
	Ruhla	Louis Bolzau Deussing Fleischmann & Co. und Eisenach in Thüringen Moritz Schlossmann Chr. Schütze Söhne Donat Thiel Kristof Triess Gebrüder Ziegler AG
	Stuttgart	C.G. Kast, Nachfabriek
	Unknown	Lux Brüder
Kingdom of Russia / Russian Empire	Riga (Latvia)	J.J. Levy
	St. Petersburg	Alexandre
Kingdom of Sweden	Smaland	Helena Sofia Isberg

COUNTRY	CITY	NAME(S)
Switzerland	Genève	Au Pacha, Fabrique de Pipes Fabrique l'Ode
United States	CA: San Francisco	S. Frohman Christian Grundel; Frederick Grundel; Julius G. Grundel Charles P. Heininger & Co. Louis Schumann Charles M. Wagner; George M. Wagner Robert Zenker Edward R. Ziesche
	CT: Hartford	L.T. Welles Co
	CT: New Haven	L.L. Stoddard
	IA: Des Moines	Des Moines Pipe Company
	IL: Chicago	I. Demuth & Co.; William Demuth & Co. Leopold Dietmann Francis Elder & Co. August Fischer Adolph Grey Henry P. Hemmerich Frederick Julius Kaldenberg; F. J. Kaldenberg; F. G. Kaldenberg; F. W. Kaldenberg & Sons; Kaldenberg & Co. Joseph Koestler M. Linkman & Co. Metzler, Rothschild & Co. John Nelson Frank Nowak Phillip Redlich Reiss Brothers
	ME: Portland	Wendel Kirsch
	MA: Boston	H. Cohen Gustav Fischer H. W. Frank F. W. Kaldenberg & Co. F. W. Steffens
	MA: Cambridge	Leavitt & Pierce
	MA: Springfield	H. H. Barnett
	MI: Detroit	Wolf Pipe Company

COUNTRY	CITY	NAME(S)
	MO: Kansas City	Fred Lederman
	MO: St. Louis	Henry Albers Co.; Rassfield & Albers
		F. J. Kaldenberg
		Max Zapf
	NJ: Camden	Edward D. Breiden
		August Fischer
	NY: Amityville	Matthew Cvitkovick
	NY: Brooklyn	Henry Berbert
		F. Blankenmeister
		Koch & Spitzer
		Aloise Leberfing
		Franz Lorenz
		Otto Shumann
		Emanuel Spitzer
		L. H. Stern, Inc.
		Herman Weinberg
		Carl and Catherine Wetzel
	NY: Cortland	Frank B. Grosse
	NY: New York City	Charles Abels & Co.
		Samuel Baldenberg
		Michael Birnbaum; William Birnbaum
		J. Blakely
		Richard J. Boiken
		Buehler & Polhaus
		Emanuel Cohn
		William Demuth & Co
		Fischer & Bendheim
		F. Fleischmann & Co.
		J. Friedrich
		I. Hamburger & Co
		C.F.A. Hinrichs
		Jobelmann & Sons
		Frederick Julius Kaldenberg
		Kaufmann Bros. & Bondy
		Joseph Koestler
		Konig & Meyer
		Carl Kutschera
		Oscar H. Lear
		Joseph Lehrkinder
		Lobe & Poggenburg
		H. Ottmann & Co.
		Pollak & Son
		Prophet & Jaburek
		Rejall and Becker (R&B)
		Em Roedel; William Roedel
		August Ruth
		Carl Schwach
	NY City (cont.)	Carl Stehr; Gustav Stehr; Charles Stehr
		Carl Weiss
		Isaac Wetzlar
	NY: Queens	William Demuth & Co.
	NY: Orchard Park	Fischer Bros.: G. A. F. ; Gustave Fischer
	NY: Rochester	J. C. Roodenburg
	OH: Cincinnati	E. Bamberger
		Chas. Keyer
	PA: Philadelphia	I. Aussprung; Aussprung & Son
		Bailey, Banks and Biddle
		C.P.F. (Consolidated Pipe Factory)
		William Demuth & Co.
		Germantown Meerschaum Pipe Co.
		Martin Holtz
		John Middleton
		N.P.W. (National Pipe Works)
		A. W. von Utassy
		George Zorn & Co.
	RI: Providence	W. Bourguignon

ANNEX B. INTERNATIONAL EXHIBITIONS AND EXPOSITIONS, 1851-1901*

Exhibition	Place, Date
The Great Exhibition of The Works of Industry of All Nations (The Crystal Palace)	London, 185
Great Industrial Exhibition	Dublin, 1853
World's Fair of Works of Industry of All Nations	New York, 1853-1854
Exposition Universelle2	Paris, 1855
International Exhibition of 1862	London, 1862
International Exhibition of Arts	Dublin, 1865
Exposition Universelle	Paris, 1867
First Annual International Exhibition	London, 1871
Second Annual International Exhibition	London, 1872
Third Annual International Exhibition	London, 1873
Weltausstellung 1873	Vienna, 1873
Fourth Annual International Exhibition	London, 1874
Exposición Internacional	Santiago, 1875
Centennial Exposition	Philadelphia, 1876
South African International Exhibition	Capetown, 1877
Exposition Universelle	Paris, 1878
Sydney International Exhibition	Sydney, 1879-1880
International Exhibition	Melbourne, 1880-1881
Internationale Koloniale en Uitvoerhandel Tentoonstelling Te Amsterdam	Amsterdam, 1883
The American Exhibition of Products, Arts, and Manufactures of Foreign Nations	Boston, 1883
International Exhibition	Calcutta, 1883-1884
World's Industrial and Cotton Centennial Exhibition	New Orleans, 1884-1885
Exposition Universelle d'Anvers	Antwerp, 1885

Colonial and Indian Exhibition	London, 1886
Jubilee International Exhibition	Adelaide, 1887-1888
Exposición Universal de Barcelona	Barcelona, 1888
Grand Concours International des Sciences et de l'Industrie	Brussels, 1888
International Exhibition	Glasgow, 1888
Centennial International Exhibition	Melbourne, 1888-1889
Exposition Universelle	Paris, 1889
International Exhibition	Kingston, 1891
Centennial International Exhibition	Melbourne, 1888- 1889
Exposition Universelle	Paris, 1889
International Exhibition	Kingston, 1891
Tasmania International Exhibition	Launceston, 1891-1892
South Africa and International Exhibition	Kimberley, 1893
World's Columbian Exposition	Chicago, 1893
Exposition Internationale d'Anvers	Antwerp, 1894
California Midwinter International Exposition	San Francisco, 1894
Tasmania International Exhibition	Hobart, 1894-1895
Exposición Centro-Americana	Guatemala City, 1897
Queensland International Exhibition	Brisbane, 1897
Exposition Internationale	Brussels, 1897
Exposition Universelle	Paris, 1900
Pan-American Exposition	Buffalo, 1901
Glasgow International Exposition	Glasgow, 1901

*John Allwood, *The Great Exhibitions*, 1977, 180-182.

ANNEX C. THE RIGHT PLACES FOR THE RIGHT STUFF

On record, as many as 200 museums, small to large, public and private, have claimed holdings of various antiquarian tobacciana artifacts, but their holdings may not always include meerschaums; and if they do, they may not always be on permanent display. For example, at one time, many American museums, such as the Metropolitan Museum of Art, New York, and the Smithsonian Museum in Washington, D.C., received donations of collections years ago, but neither museum has its collection on public display today. Hence, this list is comprised of all those museums and retail shops that are certain to have their white goddesses permanently in open display for public viewing.

Austria	Österreichisches Tabak Museum Mariahilfer Strasse 2 1070 Vienna
Belgium	Tabaks Museum Koestraat 63 8670 Wervik
Canada	Julius Vesz, Pipemaker Royal York Hotel 100 Front street West Toronto, Ontario, M5J 1E3
Denmark	W. Ø. Larsen Museum and Library 9 Amagertorv 2000 1160 Copenhagen
England	The Alfred Dunhill Museum 50 Jermyn Street London, SW1Y 6LX
France	Denise Corbier, Pipes Anciennes 3, rue de l'Odéon 75006 Paris
	Dominique Delalande Louvre des Antiquaires 2, Place du Palais Royal 75001 Paris
	Musée Galleria de la SEITA 12, rue Surcouf 75007 Paris

	Musée de la Pipe et du Diamant 45, rue de Pré 39200 Saint-Claude (Jura)
Germany	Deutsches Tabak- und Zigarrenmuseum Fünfhausener Strasse 12 32257 Bünde
	Deutsches Tabakpfeifen-Museum Valentin-Ratgeber-Haus 97656 Oberelsbach
	Heimat-Museum Ruhla Obere Lindenstrasse 29/31 99842 Ruhla
	Museumszentrum Lorsch Nibelungenstrasse 35 64653 Lorsch
	Oberrheinisches Tabakmuseum 7638 Mahlberg
	Piroska Osskó Lierstrasse 19 80639 Munich
Japan	Tobacco & Salt Museum 16-8 Jinnan 1-Chome Shibuya-ku Tokyo 150
Netherlands	Niemeyer Nederlands Tabacologisch Museum Brugstraat 24-26 9711 HZ Groningen
Spain	Pozito Preciados 1 Madrid 13

Switzerland

Au Boa Fumant
12, rue des Montbrillant
1201 Geneva

Musée de la Pipe et Objets du Tabac
Rue de l'Académie 7
1005 Lausanne-Cité

United States

California:
The Tinder Box
2729 Wilshire Boulevard
Santa Monica, 94105

Connecticut:
The Owl Shop
268 College Street
New Haven, 06510

Illinois:
Iwan Ries Company
17 South Wabash Avenue
Chicago, 60603

Louisiana:
Ye Olde Pipe Shoppe
306 Chartres Street
New Orleans, 70130

REFERENCES

BOOKS

Allwood, John. *The Great Exhibitions*. London, England: Cassell & Collier Macmillan Publishers, Ltd., 1977

An Old Smoker. *Tobacco Talk*. Philadelphia, Pennsylvania: The Nicot Publishing Company, 1894

Anon. *An Arm-Chair In The Smoking-Room: Or Fiction, Anecdote, Humour, and Fancy For Dreamy Half-Hours. With Notes On Cigars, Meerschaums, and Smoking, From Various Pens*. London, England: Stanley Rivers and Co., 1868

Armero, Carlos. *Antique Pipes. A Journey Around A World.* Madrid, Spain: Tabapress, 1989

Bock, C. *Der echte Wiener Meerschaumkopf als Schmuck des Tabakrauchers. Theoretisch-praktische Anweisung, Meerschaumköpfe verschiedener Form sicher und gut anzurauchen, wie überhaupt Pfeifen gehörig zu behandeln.* Wien/Leipzig: Haas, *1843*

Bozzini, Giuseppe. *Le Pipe.* Vetrina del Collezionista. Legano (Mi), Italy: EdiCart srl, 1995.

Constantin. *Biographisches Lexicon des Kaisertums Österreich.* Vienna, Austria: 1874

Cope Brothers & Company. *Pipes and Meerschaums. Part The Third. English and European Pipes.* Cope's Smoke Room Booklet Number Twelve. Liverpool, England: Cope's Tobacco Plant, 1896

Cudell, Robert. *Das Buch vom Tabak.* Cologne, Germany: Verlag Haus Neuerburg, 1927

Cundall, J.W. *Pipes and Tobacco. Being A Discourse on Smokers and Smoking.* London, England: Greening & Co., Ltd., 1901

Fairholt, Frederick W., F.S.A. *Tobacco: Its History and Associations: Including An Account of The Plant and Its Manufacture; With Its Modes of Use in All Ages and Countries.* London, England: Chapman and Hall, 1859

Findling, John E. (ed.). *Historical Dictionary of World's Fairs and Expositions, 1851-1988.* Westport, Connecticut: Greenwood Press, 1990

Heward, Edward Vincent. *St. Nicotine of The Peace Pipe.* London, England: George Routledge & Sons, Ltd., 1909

Hiscox, Gardner D., M.E. (ed.). *Henley's Twentieth Century Book of Formulas, Processes and Trade Secrets.* New revised and enlarged edition. New York, New York: Books, Inc., 1957

Innes, Brian (comp.). *The Book of The Havana Cigar.* London, England: Orbis Publishing Limited, 1983

Koppa, G. *Über Meerschaumwaaren.* Pirna, Germany: 1873

Levárdy, Ferenc. *Our Pipe-Smoking Forebears.* Translated from Hungarian by Andrew C. Rouse and Imre Eliás. Budapest-Pécs, Hungary, and Velburg, Germany: Druckhaus Oberpfalz, 1994

Meyer, Alois. *Ueber Meerschaum-, Bernstein- und Drechslerwaaren in der Weltausstellung 1873.* Vienna, Austria: 1874

Museum of Tobacco Art and History. *Meerschaum Masterpieces: The Premiere Art of Pipes*. Nashville, Tennessee: December, 1990
Penn, W. A. *The Soverane Herbe. A History of Tobacco*. London, England: Grant Richards, 1901
Ram, Sidney P. *How To Get More Fun Out of Smoking*. Chicago, Illinois: Cuneo Press, Inc., 1941
Ramazzotti, Eppe, and Bernard Mamy. *Pipes et Fumeurs de Pipes. Un Art, des Collections*. Paris, France: Editions Sous le Vent, 1981
Rapaport, Benjamin. *A Complete Guide to Collecting Antique Pipes*. Exton, Pennsylvania: Schiffer Publishing Ltd., 1979. Second edition, 1998
______. *Museum of Tobacco Art & History Guide Book*. Nashville, Tennessee: 1996
Raufer, G.M. *Die Meerschaum- und Bernsteinwaaren-Fabrikation. Mit einem Anhange über die Erzeugung hölzerner Pfeifenköpfe*. Vienna, Pest, Leipzig: A. Hartleben's Verlag, 1876
Reemtsma, H.F. & Ph. F. *Tabago. A Picture-Book of Tobacco and The Pleasures of Smoking*. Hamburg, Germany, 1960
Reineke, Hugo (verleger). *Die Meerschaum-Industrie. Inaugural-Dissertation*. Wesermünde-Lehe: Friedrich Reimann, 1930
Scott, Amoret, and Christopher Scott. *Tobacco and The Collector*. London, England: Max Parrish, 1966
Siluriensis, Leolinus (pseud.). *The Anatomy of Tobacco: or Smoking Methodised, Divided, and Considered After A New Fashion*. London, England: George Redway, 1884
Tomas, J.A., Mechanicus. *Praktische Anleitung meerschaumene Pfeifenköpfe zu verfertigen, ächte von unächten zu unterscheiden, nebst den Vortheilen, solche in Wachs und Talg zu sieden, anzurauchen und auch den schlechtern Massen die angerauchte Farbe zu geben*. Erlangen, Germany: Johann Jakob Palm, 1799
Tomasek, J.N. *Die Pfeifen-Industrie auf der Höhe jetziger Zeit, oder die Fabrikation der Tabaks- und Cigarrenpfeifen aus Meerschaum und allen Arten Holz*. Weimar, Germany: Voigt, 1878
Thon, Christian Friedrich Gottlob. *Gründliche und vollständige, auf richtige Erfahrungen gestützte Anleitung, nicht allein alle Arten meerschaumener, sondern auch hölzerner Pfeifenköpfe fabrikmässig herzustellen...Nebst einer geprhpften Anweisung, meerschaumenen Pfeifenköpfe anzurauchen, ihnen beliebige Farbe zu geben, beschmutzte zu reinigen*. Weimar, Germany: Voigt, 1833
United States Centennial Commission. *International Exhibition, 1876. Official Catalogue*. Philadelphia, Pennsylvania: John R. Nagle and Company, 1876
Voges, Ernst (comp. and ed.). *Tobacco Encyclopedia*. Mainz, Germany: Tobacco Journal International, 1984
Werner, Carl Avery. *Tobaccoland*. New York, New York: The Tobacco Leaf Publishing Company, 1922
Ziegler, Dr. Alexander. *Zur Geschichte des Meerschaums mit besonderer Berücksichtigung der Meerschaumgruben bei Eski Schehr in Kleinasien und der betreffenden Industrie in Ruhla in Thüringen*. Dresden, Germany: Carl Höckner, 1878

PERIODICALS

Becker, Nancy. "Meerschaum Masterpieces," *Eagle*, Spring 1991
"Buffalo Pipe Carver for Half Century Upholds Traditions of Famous Family," *The Buffalo Times*, March 24, 1934
Cope Brothers & Company. "Exhibition Pipes at South Kensington," *Cope's Tobacco Plant*, No. 18. Vol. I, September 1871
______. "Meerschaum," *Cope's Tobacco Plant*, No. 59, Vol. I, February 1875
______. "Meerschaum II," *Cope's Tobacco Plant*, No. 60, Vol. I, March 1875
______. "Meerschaum III," *Cope's Tobacco Plant*, No. 61, Vol. I, April 1875

______. "Meerschaum IV," *Cope's Tobacco Plant*, No. 63, Vol. I, June 1875
______. "Meerschaum V," *Cope's Tobacco Plant*, No. 64, Vol. I, July 1875
______. "Meerschaum VI—Conclusion," *Cope's Tobacco Plant*, No. 66, Vol. I, September 1875
______. "The Pipes of Various Races," *Cope's Tobacco Plant*, No. 10, Vol. I, January 1871
______. "Tobacco-Pipes: With Special Reference to The Collection in The London International Exhibition, 1873," *Cope's Tobacco Plant*, No. 44, Vol. I, November 1873
Copeland, Ed. "Determining Pipe Value," *Pipe Lovers*, Vol. IV, No. 7, July 1949
______. "Evaluating Meerschaums," *Pipe Lovers*, Vol. IV, No. 1, January 1949
Cosand, Gary. "Guidelines for Evaluating Meerschaums," *The Antiques Journal*, January 1975
Crutchfield, James A. "The United States Tobacco Company's Museums," *The Magazine Antiques*, Vol. CXXX, No. 3, September 1986
Devilbliss, Philip. "Pictorial Essay on Meerschaum Pipes," *The Antiques Journal*, July 1973
Duncan, C. S. "Coloring A Meerschaum," *Pipe Lovers*, Vol. V, No. 2, February 1950
Duncan, E. Reid. "Mysteries of Meerschaum," *Pipe Lovers*, Vol. IV, No. 1, January 1949
Everett, Dale. "He Demands Perfection," *Pipe Lovers*, Vol. V, No. 3, March 1950
______. "For The Pipe Collector," *The American Smoker*, Vol. 5, No. 10, October 1950, and Vol. 6, No. 1, January 1951
Fresco-Corbu, Roger. "How to Smoke An Economical Cigar," *Country Life*, April 1, 1965
______. "Mystic Meerschaum," *Art & Antiques Weekly*, January 20, 1973
______. "The Era of The Meerschaum Pipe," *Country Life*, November 10, 1960
Friedman, Frances. "Meerschaum Pipes," *NEAA News*, May 1984
Fusco, Anthony. "The Queen of Pipes," *The Antiques Journal*, February 1978
Grenville, Arthur. "Meerschaum as a Pipe For The Connoisseur," in Gordon West (ed.), *All About Pipes and Pipe Tobaccos. A Handbook For Tobacconists*, London, 1953
Harte, J. "Famous Fischers," *Pipe Lovers*, Vol. III, No. 3, March 1948
Hillier, Bevis. "Pipe Dreams," *Punch*, August 31, 1983
Hubbard, Clarence T. "Classic Pipes," *The Antiques Journal*, August 1964
Kolpin, Ed, Jr. "How To Color A Meerschaum," *Pipe Lovers*, Vol. II, No. 3, March 1947
Leighton, George. "Emperor's Pipes," *Pipe Lovers*, Vol. III, No. 3, March 1948
Littledale, Harold. "Meerschaums," *U.S. Tobacco Review*, Fall 1979
_____. "Meerschaums. The Aristocrat of Pipes," *Spinning Wheel*, July-August 1981
"Meerschaum and Amber," *The Scientific American*, XV:II, September 8, 1866
"Meerschaum Pipes: How They Are Made, and The Science of Coloring Them Properly," *The New York Times*, July 27, 1874
Melinsky, Denise. "A Passion For Pipes," *Smokeshop*, August 1990
Moore, Thomas. "Meerschaum Pipes," *Pipe Lovers*, Vol. I, No. 1, January 1946
Morris, Fritz. "Making of Meerschaums," *Technical World*, April 1908
Morrison, James. "The World's Largest Meerschaum Pipe," *Pipe Lovers*, Vol. II, No. 2, February 1947
Musgrave, William. "Famous Meerschaum," *Pipe Lovers*, Vol. IV, No. 6, June 1949
O'Connor, Larry. "A Remembrance of Pipes Past," *Pipes and Tobaccos*, Vol. 1, No. 2, Spring 1996

Piercy, Carl A. "Carved Meerschaums Plus Historical Pieces Make An Interesting Collection," *Pipe Smokers Review*, Vol. 1, No. 1, May 1952

Ramazzotti, Giuseppe. "In a Harem of Beauties," *Pipe World* (North American edition), No. 1, May 1970

______. "Light A Meerschaum...Light As Sea Foam," *Pipe World* (North American edition), No. 1, May 1970

Rapaport, Benjamin. "About The Origin of Meerschaum Pipes," *Smokeshop*, May 1986

______. "More Than Blowin' Smoke! Collecting Antique Tobacco Pipes," *Antiques & Collecting Hobbies*, January 1989

______. "Pipes of Our Presidents," *Antiques & Collecting Magazine*, November 1994

______. "Spotlight on Meerschaum Pipes," *Smokeshop*, February 1984

______. "The Magnificent Memorable Meerschaum," *The Antique Trader Weekly*, June 18, 1974

______. "Tobacciana At Auction 1990: It Was A Very Good Year!," *The Compleat Smoker*, Vol. 1, No. 3, Spring 1991

"Remembrance of Things Past: Nashville Museum Showcases History of Tobacco Use," *U.S. Tobacco Review*, Vol. 1, No. 3, Third Quarter 1985

Rich, Tim. "So You Want To Be A Pipe Collector," *Pipe, The Worldwide Pipe Smoker's Magazine*, Vol. 1, First Semester 1995

Rupp, Herbert. "Viennese Meerschaum," *Pipes and Tobaccos*, Vol. 1, No. 1, Winter 1996

Schiff, David. "Ornate Meerschaums Tell Stories of Life Since 17th Century," *Smokeshop*, May 1978

Shulsinger, Stephanie Cooper. "Meerschaum Pipes," *Relics*, December 1975

Spencer, C. Bruce. "The Emperor's Pipe. It Deserves To Be Seen!," *Pipe Smoker*, Vol. 3, No. 4, Fall 1985

White, Clair. "How Meerschaums Are Made," *Pipe Lovers*, Vol. IV, No. 7, July 1949

Wilford, Sarah. "10,000 Pipes — at Every Price," *Long Island Press*, January 24, 1936

Wingert, Robert. "Thirty Years A Collector," *Pipe Lovers*, Vol. II, No. 12, December 1947

OTHER SOURCE MATERIAL

Edinburgh Museum of Science and Art. Science and Art Department. *Guide to The Loan Collection of Objects Connected With The Use of Tobacco and Other Narcotics*. Edinburgh, Scotland: Neill and Company, 1880

B. Linger and Company of Vienna, Austria and London. *Illustrated Catalogue. Wholesale and Export Only*, ca.1900

Salmon & Gluckstein, Limited. *Illustrated Guide For Smokers*. London, England, ca. 1899

George Zorn & Co. *Pipes, Matches, Walking Canes, and Playing Cards. Illustrated Catalogue*, 3rd edition. Philadelphia, Pennsylvania, ca. 1886-87

______. *Pipes & Smokers Articles. Illustrated Catalogue*, 5th edition. Philadelphia, Pennsylvania, ca. 1892. Privately reprinted. Bel Air, Maryland: S. Paul Jung, 1989

PRICE GUIDE

The price ranges in this guide are current-market values, and current-market value is determined by several parameters, each of which contributes toward determining the price of antique meerschaums. The first parameter is present condition, e.g., there are no breaks, fissures, cracks, chips, or flakes to either the body or the mouthpiece. Next, the material must be original: the meerschaum has not been repaired or restored to conceal prior damage; the amber mouthpiece has not either been fused together with a metal mount (shank band) or replaced by some artificial material; and, of much less importance, the fitted case is present. (If the pipe-maker fitted a metal mount to the shank, it was almost always silver or sterling and distinguishable by a standard hallmark, a manufacturer's symbol, or a die stamp, and, just as often, the mount was ornately chased.) Overall size of the object is important, but it is a very difficult parameter to gauge in the valuation. As a general rule, large is assigned greater value than small; but, often, small exhibits more precise and intricate detail than large, in which instance, small may have greater attendant value than large. The last parameter, motif, is, by far, the most difficult to judge since years of collecting is the experiential foundation of knowledge that provides the valuator the ability to distinguish the common from the rare. As is the case with many other fields of antique collecting, it is almost impossible for the uninitiated to know whether a particular motif is ubiquitous or unusual, run-of-the-mill or rare. Thus, assigning a part of the value to motif is, without question, exceptionally difficult and remains the subject of much inspired and impassioned deliberation among even the most knowledgeable collectors.

The vast majority of the pipes and cheroot holders illustrated in this book are in a mint, near-mint, or pristine state, unless otherwise noted in the captions. These price ranges are my best approximation of their current-market value; they were derived by (1) assigning a U.S. value to each object based on its condition and authenticity, size, and motif, and then (2) adjusting each value slightly upward to reconcile with/reflect the price structure for similar items encountered on the Continent. (Today, in almost every instance, the price for an antique meerschaum in Europe of comparable size, condition, execution, and motif to one offered on the open market in the United States is more costly by *at least* a factor of 1.5.)

Buyers should note that antique dealers and auction houses often have sold fine-quality antique meerschaum pipes and cheroot holders like those appearing in this book for much more. Conversely, sellers should note that many collectors are able to purchase exquisite meerschaum pipes and holders for far less than the prices cited herein. In brief, then, I offer a counsel to both sellers and buyers who may now govern their decisions on value by strict application of this price guide: *caveat emptor* and *caveat vendor.*

The Price Guide has three columns. The page number is in the left-hand column; the illustration is in the center column; and the price range in U.S. dollars is in the right-hand column. When more than one illustration appears on a page, the following letter code indicates the position of each illustration: T=top; C=center; B=bottom; R=right; and L=left. Hence, TC=top center, and BR=bottom right. If the illustration contains several items, then the code applied to that illustration is as follows: TT=top, top; TC=top, center; and TB=top, bottom.

Cover		3000-3500	38		500-700	63	B	2500-3500	69	B	350-500
4		20000-25000	41		500-700	64	T	2500-3500	70	T	350-500
15		400-500	42		4000-5000	64	B	3000-4000	70	C	350-500
16		400-550	43		500-700	65	T	1500-2200	70	B	350-500
17		500-700	44		3000-4500	65	B	2500-3500	71	T	250-350
18	T	700-1000	54		0	66		300-400	71	C	400-500
18	B	800-1000	59	T	8000-10000	67	T	250-350	71	B	350-450
33		750-900	60	T	700-900	67	CT	200-300	72	T	450-600
34	T	600-800	60	C	450-650	67	CB	250-350	72	C	450-650
34	B	700-900	60	B	250-350	67	B	250-350	72	B	250-350
35	T	1200-1600	61	T	1200-1600	68	T	400-600	73	T	600-800
35	B	600-900	61	B	1500-1800	68	C	300-400	73	C	800-1200
36		3500-4500	62	T	600-900	68	B	350-450	73	B	800-1100
37	BT	300-350	62	B	6000-8000	69	T	500-650	74	T	2000-3000
37	BB	350-450	63	T	1700-2000	69	C	350-500	74	B	350-450

75	T	250-300
75	C	300-350
75	B	250-350
76	T	300-400
76	CL	300-400
76	CR	350-450
76	B	700-900
77	T	750-1000
77	C	700-850
77	B	600-800
78	T	700-800
78	B	850-1000
79	T	1200-1600
79	B	1200-1500
80	T	700-900
80	C	900-1200
80	B	2000-2500
81	T	1200-1600
81	C	800-1000
81	B	700-850
82		2500-3200
83	T	500-700
83	C	500-650
83	B	400-500
84		400-500
85	T	350-500
85	C	250-350
85	BT	250-350
85	BB	250-350
86	T	400-500
86	C	250-300
86	B	150-200
87	TT	250-350
87	TC	250-350
87	TB	250-350
87	C	400-500
87	B	450-550
88	T	300-400
88	C	900-1000
88	B	700-900
89	T	700-900
89	C	450-600
89	B	550-700
90	T	1200-1500
90	CT	300-400
90	CB	450-550
90	B	400-500
91	T	400-500
91	C	450-550
91	B	450-550
92	T	800-900
92	C	400-500
92	B	450-550
93	T	900-1100
93	C	700-900
93	B	800-1000
94	T	850-1000
94	B	3500-4500
95	T	3000-3500
95	B	1500-2200
96	T	600-750
96	C	600-750
96	B	600-700
97	T	600-750
97	C	600-750
97	B	600-800
98	T	500-650
98	CT	350-400
98	CB	350-400
98	B	350-400
99	T	400-500
99	C	350-500
99	B	400-550
100	T	700-850
100	C	1100-1400
100	B	900-1100
101	T	1400-1800
101	C	800-1000
101	B	750-900
102	T	750-900
102	C	400-500
102	B	400-500
103	T	400-500
103	C	400-500
103	B	450-550
104	T	600-800
104	C	1000-1400
104	B	800-1000
105	T	800-1000
105	C	800-1000
105	B	800-1000
106	T	800-1100
106	C	700-900
106	B	500-700
107		400-500
108	TL	350-450
108	TR	350-450
108	B	2200-2500
109	T	1200-1500
109	C	600-700
109	B	400-500
110	T	900-1200
110	C	700-900
110	B	600-800
111	T	1200-1500
111	B	2500-3500
112	T	1500-2000
112	C	900-1200
112	B	1200-1500
113	T	900-1300
113	CL	350-450
113	B	500-700
114	T	450-500
114	C	400-500
114	B	350-450
115	T	350-500
115	B	250-350
116	T	300-400
116	C	250-350
116	B	800-1000
117		400-600
118	TT	250-350
118	TB	250-350
118	C	250-350
118	B	200-300
119	T	300-350
119	C	450-550
119	B	450-550
120	T	450-650
120	C	400-600
120	B	250-350
121	T	400-550
121	C	700-900
121	B	350-450
122	T	300-400
122	C	300-400
122	B	600-800
123	T	400-550
123	C	250-350
123	B	350-450
124	T	400-500
124	C	250-350
124	B	250-350
125	T	350-450
125	C	350-450
125	B	500-650
126	CT	350-400
126	CB	400-500
126	B	350-450
127	T	400-450
127	B	2200-2700
128		350-450
129	TL	350-450
129	TR	350-450
129	C	400-500
129	BT	250-350
129	BB	250-350
130	TT	200-300
130	TB	250-300
130	C	400-450
130	B	750-900
131	TT	200-300
131	TB	200-300
131	CL	200-300
131	CR	300-400
131	B	500-600
132	T	300-400
132	CT	150-200
132	CC	150-200
132	CB	150-200
132	BT	150-200
132	BC	150-200
132	BB	150-200
133	T	250-350
133	CLT	150-250
133	CLB	200-300
133	CR	250-350
133	BT	250-350
133	BB	350-450
134	T	350-500
134	CT	350-450
134	CB	300-400
135	B	300-400
135	T	250-350
135	C	300-400
135	B	1000-1300
136	TT	200-300
136	TB	200-300
136	C	250-350
136	BT	200-300
136	BB	200-300
137	TT	200-300
137	TB	250-350
137	C	350-450
137	B	250-350
138	T	400-500
138	CT	250-300
138	CC	250-350
138	CB	300-400
138	B	350-500
139	T	350-450
139	C	250-350
139	B	450-600
140	T	450-650
140	B	1800-2400
141	T	200-300
141	C	900-1200
141	B	800-1200
142	T	3000-4500
142	C	2000-2500
142	B	1600-2000
143	T	1000-1400
143	C	500-750
143	B	350-450
144	TL	350-500
144	TR	500-750
144	B	3000-3500
145		350-500
146	T	350-500
146	C	350-500
146	B	600-800
147	T	850-1100
147	C	600-900
147	B	600-800
148	T	400-550
148	C	1000-1300
148	B	700-900
149	T	1300-1700
149	C	300-400
149	B	500-800
150	T	250-350
150	C	250-350
150	B	300-400
151	T	400-500
151	C	300-450
152	T	450-650
152	B	400-600

INDEX

A
Achmed III, Sultan, 16
Adams, John, President, 22
Adler, Budapest, 33
Aesop, 107
Ainé, Marot, 25
A la Sans Rivale, 25
A l'Indien, 25
Amber, 25, 35, 39, 40, 44, 49, 53
American Institute Fair, 43
American Museum of Natural History, 45
American Tobacco Company, 23, 45
Andrássy, Count, 16, 17, 23
Aphrodite, 79
Appraisal of meerschaums, 47
Archilochaus, 3
Arents, George Jr., 45
Athena, 112
Augustus Frederick, Duke of Sussex and Earl of Inverness, 8
Austrian Tobacco Museum, 45
Aux 100,000 Pipes, 25

B
Bacchus, 36
Bakelite, 39
Balaklava, charge of, 42
Battle of Bunker Hill pipe, 24
Battle of Sadowa, 2
Battle of Trafalgar, 60
Bedouin, 48, 49
Bedrossian, 21
Berini, 17
Bernini, Giovanni Lorenzo, 17
Birmingham and Midland Institute, Birmingham, England, 45
Bismarck-Schonhausen, Karl Otto Furst von, 93, 137
Bragge, William, F.S.A., 8, 45
Brancusi, 9
Br'er fox, 107
Brennan, Jane, 48
Briar, 38
Brittania, 60
Bruce Museum, Greenwich, Connecticut, 24
Budapest, 16

C
Canada, 44
Cellini, 9
Celluloid, 39
Centennial Exposition, 1876, 41, 44
Cheroot, 36
Cheroot holder, 8, 36, 37, 48
Cherubs, 56, 117
Chinese amber, 39
Christie's South Kensington, 52, 55
Cigar, 8, 36, 37
Cigarette, 8, 37
Cigarette holder, 8
Cigar holder, 15
Cigar-store Indians, 22
Columbia pipe, 44
Columbus Landing in America pipe, 22, 44, 45
Composition, 39
Constantinople, 37
Crimean War, 37
Crystal Palace, London, 41, 80
Cundall, J. W., 39
Cupid, 56
Czapek, 26, 55, 60, 85

D
da Cortona, Pietro, 17
Debrecen, 31, 33, 34
Delrin, 39
Demuth, Leopold, 21
Demuth, Louis, 21
Demuth, William, 21, 23, 29, 44, 45, 55
de Watteville, Baron Oscar, 8
Dewey, George, Admiral, 24
Diana and the hunt, 38, 48, 56, 109
Doolittle, Amos, 24
Dragon's blood, 14
Dresden, 34, 43
Drucquer & Sons, Ltd., 52
Duke of Sussex, 8
Dumas, Alexandre, 114
Duncan, E. Reid, 17
Dunhill, Alfred, Pipes Ltd., 2

E
E-commerce, 51
Earl of Inverness, 8
Edinburgh Museum of Science and Art, 45
Ehrlich, David P., & Company, 24
Elizabeth, Princess of Austria, 80
Erotic pipes, 56, 143
Exposition Universelle, 43, 45, 59

F
Fairholt, Frederick W., F.S.A., 36
Fatherland Antique Vienna Meerschaum, 53
Fischer, Arthur C., 23
Fischer, August G., 23, 43
Fischer, Gustav, Jr., 24, 54, 55
Fischer, Gustav, Sr., 24, 30, 55
Fischer, Gustave A., 23, 24, 55
Fischer, Otto, 23
Fischers of Massachusetts, 22, 24
Fischers of New York, 22, 23
Frank, S. M., & Company, 23
Frederick the Great (Frederick II, King of Prussia), 34

G
Gambrinus, 114, 115
Ganymede, 112
Garfield, James A., President, 22
German antique pipe, 53
Gesteckpfeife, 35
Gibson girl, 47, 48
Goas, Messrs. Charles, and Co., 42
Goethe, 114
Goltsche, Alexandre, 25
Goltsche, Auguste, 25
Goltsche, Charles, 25
Goltsche, François (Frantz), 25
Goltsche, Roger, 25
Great Exhibition, 1851, 41
Guyot, Gilbert, 24, 25, 26
Guyot, Henri, 25
Guyot, Philippe, 25

H
HMS Victory, 60
Half and Half Collection, 23, 45
Hard rubber, 39
Harris, Joel Chandler, 107
Hartmann, 26
Hartmann & Eidam, 42
Heide, John, F. H., 8
Held, M., 41
Hen, Edward, 22
Hiess, Franz, 42
Historical Museum of Southern Florida, 45
Holy Roman Empire, 19
Hoover, Herbert, President, 22, 23
Horn, 39
Hudson's Bay Company, 53

Hungarian lap-style pipes, 16, 17, 39
Hungarian National Museum, 16
Hydrous magnesium silicate, 9

I
Internal Revenue Service, 48
Internet, 51

J
John III Sobieski, King of Poland, 16

K
Kaldenberg & Son Company, 20, 26, 42, 44
Kaldenberg, F. J., 27, 28, 43
Kaldenberg, F. W., 21
Kalmasch (Kalmas), 31, 33, 34
Kapp & Peterson, 12
Koppa, Dresden, 33-34
Kovács, Karl (Károly Kovács, or Karol Kowates), 16, 17
Kutschera , 26

L
Le Brigand au repos, 17, 62
Leda and the swan, 109
Leipzig, 21
Lemgo, 19
Levárdy, Ferenc, 17
Linger, B., Company, 12, 13, 14
Linseed oil, 14
London, 21, 43
London (Kensington) International Exhibition, 41
L'Oriental, 25
Lucite, 39

M
Macbeth and Banquo, 42, 43, 44
Maillol, 9
McKinley, William, President, 23, 24
Meerschaum, block, 12
Meerschaum, calcinate, 12
Meerschaum, carving of, 13
Meerschaum chips, 12,
Meerschaum, coloration of, 14
Meerschaum, compressed, 12
Meerschaum, dating of, 31
Meerschaum, definition, 11
Meerschaum, discovery of, 16
Meerschaum, fakes and forgeries, 53
Meerschaum, hard, 12
Meerschaum, imitation, 12, 13, 19, 53
Meerschaum-masse, 12
Meerschaum, medium, 12
Meerschaum, metamorphosis of, 31
Meerschaum, mock, 12
Meerschaum, pseudo-antique, 54
Meerschaum, soft, 12
Meerschaum, Viennese, 12
Mephistopheles, 48, 132
Mercury, 118
Mere-sham, 12
Michelangelo, 9, 49
Montgomery-Ward, 53
Moore, 9
Munich, 43
Museum of American Folk Art, 45
Museum of Tobacco Art and History, 3, 24, 46
Mythology, 107

N
Naiads, 56, 117
Naïve pipes, 56, 145
Napoleon III, 44
Nelson, Horatio, Admiral, 60
Neptune, 35
Netsukes, 8
New York Public Library, 45
Nowatry, M. A., 42
Nubian, 48, 49, 83, 86, 94, 132
Nürnberg, 19, 41

P
Palais Royal, Paris, 25
Pan, 48, 110
Pan American Exposition, 1901, 23, 43
Paris, 21, 25, 43, 51
Peabody Museum of Natural History, 45
Pest, Hungary, 16
Petersburg, 37
Pipe, definition of, 10
Pipe Lovers, magazine, 17
Pornographic pipes, 56
Poseidon, 35, 111
Prince of Saxony, 23
Provenance of meerschaums, 39
Puck, 126
Puget, Louis Pierre, 17
Putti, 56, 117

R
R.J. Reynolds Company, 45
Rape of the Sabine Women, 61
Rákcózi (Rágóczy), 31, 33
Raufer, G. M., 21
Redolite, 39
Reynard the Fox (Reineke Fuchs), 115, 116
Robb, Samuel Anderson, 22
Rochester Museum of Arts and Sciences, 45
Roedel, 26
Rottenstone, 14
Ruhla, 19, 43

S
Saint-Gaudens, 9
Salmon & Gluckstein, 32
Schnally, 26
Schulte, D. A., 23
Schwager, Joseph, 35
Sears Roebuck, 53
Seraphim, 56, 117
Seven-Year War, 19
Shakespeare, William, 42, 43, 44, 126
Sheffield, England, 45
Soapstone, 9
Sommer Frères, 26
Sotheby Parke Bernet, 52
Spermaceti, 13
Stearine, 13
Steinbeck, John, 84
St. George and the dragon, 41, 107, 108
Suggestive pipes, 56, 145

T
The Netherlands, 44
Theseus, 61, 111
The Three Graces, 65
The Three Musketeers, 114
Tomas, J. A, 17
Torpedo, 37
Trebitsch, A., 42
Turkey, 11, 39, 54,
Turkish (meerschaum) pipes, 32, 53

U
Uncle Remus, 107
Uncle Sam, 92
U.S. International Exhibition of 1876, 41
USS Maine, 58
U.S. Tobacco Museum, 48

V
Valentine Museum, 23, 45
Valuation of meerschaums, 47
Venus, 48, 79
Vienna, 16, 19, 21, 43, 49, 51
Vienna Academy of Design, 24
Vienna Exposition 1873, 54, 55
von Moltke, Count (Graf) Helmuth Karl Bernhard, 91

W
WDC, 22
Wagner, Richard, 113
Warren, Joseph, General, M.D., 24
Waterloo, 58
Weiss, G., Company, 21
Weiss, Philipp & Söhne, 26
Weiss (Weisz), Budapest, 33
Wellesley, Arthur, Iron Duke, Duke of Wellington, 34
Weltausstellung Wien, 55
Wickenburg, Carl, 34
Wiegleb, Johann, 12
World's Columbian Exposition, 44
World's Industrial and Cotton Centennial Exhibition, 1884-1885, 44
Wright, David R., 3, 46

Y
Yale Center for British Art, 45

Z
Zeus, 111, 112
Zorn, George & Company, 12, 25